Malware - Detection and Defense

Edited by Eduard Babulak

Published in London, United Kingdom

Malware - Detection and Defense
http://dx.doi.org/10.5772/intechopen.105262
Edited by Eduard Babulak

Contributors
Iyas Alodat, Arun Warikoo, Cristiane M. Batistela, José Roberto C. Piqueira, Joshua Ojo Nehinbe, Benjamin Okorie Ajah, Obinna J. Eze, Chukwuemeka Dominic Onyejegbu, John Thompson Okpa, Liqaa Nawaf, Nisha Rawindaran, Vibhushinie Bentotahewa, Edmond Prakash, Ambikesh Jayal, Chaminda Hewage, Daniyal Mohammed N. Alghazzawi

First published in London, United Kingdom, 2023 by IntechOpen
IntechOpen is the global imprint of INTECHOPEN LIMITED, registered in England and Wales, registration number: 11086078, 5 Princes Gate Court, London, SW7 2QJ, United Kingdom

British Library Cataloguing-in-Publication Data
A catalogue record for this book is available from the British Library

Additional hard and PDF copies can be obtained from orders@intechopen.com

Malware - Detection and Defense
Edited by Eduard Babulak
p. cm.
Print ISBN 978-1-83768-444-1
Online ISBN 978-1-83768-445-8
eBook (PDF) ISBN 978-1-83768-446-5

Meet the editor

Professor Eduard Babulak is a global scholar, consultant, engineer, and polyglot. He is a published author and served as chair of the Institute of Electrical and Electronics Engineers (IEEE) Vancouver Ethics, Professional and Conference Committee. He has been an invited speaker at numerous universities, including the University of Cambridge, Massachusetts Institute of Technology, Purdue University, Yokohama National University, University of Electro-Communications, Shanghai Jiao Tong University, Sungkyunkwan University, Penn State, Czech Technical University, University of the West Indies, Graz University of Technology, and other prestigious academic institutions worldwide. He serves as an editor-in-chief, associate editor-in-chief, co-editor, and guest editor for many publications. He communicates in sixteen languages. Dr. Babulak's biography has been cited in the *Cambridge Blue Book*, the *Cambridge Index of Biographies*, *Stanford Who's Who*, and *Who's Who in the World and America*.

Contents

Preface

Given the current dynamics of Internet innovation, research, and development, the proper provision of cyber security has become essential for governments, businesses, industries, and academic institutions worldwide. Malware detection and defense continue to be most challenging for network administrators and cyber critical infrastructures across the globe. Hardware and software vulnerabilities are often exploited via malware, which may compromise cyber system integrity and business continuity. The evolution of malicious malware drives the evolution of new malware detection and defense in support of the resilience and reliability of cyber systems.

The most common cyberattacks are caused by malware (malicious software). Common malicious software tools today include viruses, spyware, worms, trojans, ransomware, rootkits, keyloggers, and more. These tools can take control of an infected computer and compromise the system's integrity, ultimately causing a shutdown. Numerous reports indicate that close to a half million cyberattacks are caused by malware. Given the computational complexity of the current Internet and cyber critical infrastructures, the number of malware has and will continue to increase dramatically.

The book provides a comprehensive overview of malware detection and protection in support of creating reliable and resilient cyber and computer security mechanisms. Each chapter brings to light the most recent theoretical and applied research findings, accompanied by practical case studies. The book is an excellent reference for cybersecurity professionals, practitioners, scholars, and students. It promotes the creation of global interdisciplinary research teams to face new cyber challenges today and in years to come.

Eduard Babulak
National Science Foundation,
Alexandria, VA, USA

Section 1

Social Impact

Chapter 1

Cybercrime: Victims' Shock Absorption Mechanisms

Obinna J. Eze, John Thompson Okpa,
Chukwuemeka Dominic Onyejegbu and Benjamin Okorie Ajah

Abstract

The development of technology creates opportunities for businesses, seamless communications and leisure activities to thrive. However, it also propels crime. In Nigeria, cyber threat continues to evolve rapidly with rising number of victims on daily bases. This necessitated the present study that examines the shock absorption mechanism of the cybercrime victims in Nigeria. The data for this study came from a variety of sources, including books, articles, essays, tabloids, and journal publications; a content analysis approach was used to evaluate the data and present using certain words, themes, concepts, or codifications. The study found that the peculiarity of cybercrime lies in the fact that the victims willingly land themselves into it without being forced to do so. It starts with what seem to be a friendly conversation and exchange of correspondences and pleasantries which turns into a scamming spree. To this end, victims are left battered and shattered, and could act irrationally against own-self before state actors set out to track the offender(s). Thus, victims of cybercrime could absorb shock by spending quality time with significant others. This enables them feel the love and moral supports from close associates, other than wallow in loneliness and isolation which can breed unpleasant stimuli.

Keywords: cyber-attacks, cybercrime, law enforcement agencies, shock absorption mechanisms, victims

1. Introduction

Cyber-attacks are growing in multiple dimensions globally. Malicious cyber activity poses a danger to public safety, national security, and economic stability. The global cyber threat continues to evolve at a rapid pace, with a rising number of people falling victims on daily bases. The development of technology creates opportunities for people, such as business and leisure activities, but also enables criminals to commit crimes [1–3]. Research conducted by Pew research center (PEW) in 2014, reveal an astonishing 40% of all adult internet users admit to experience cyber victimization of different variants [4]. Most often, cyber-attacks such as hacking, phishing, business email compromise (BEC) malware attacks, password attacks, man-in-the-middle attacks, insider threats, and crypto-jacking are the most frequently suffered by victims [5]. Cybercriminals have access to a wide variety of psychological manipulation techniques. For instance, phishing emails are the most typical means through which hackers distribute ransomware. Fake emails are also used by hackers to deceive victims into opening dangerous attachments or clicking on hazardous links. Cybercriminals also exploit the natural desires of humans to trust others to send unsolicited electronic mails to unsuspecting victims, as though they originated from legitimate sources [6–8].

The betray of trust is very common technique used by dating fraudsters, they engage in the purposeful creation of trust with their victim, often over a period of weeks or months, with the goal of betraying them after extorting money from them. To learn later that a relationship that seems to be based on openness, closeness, and trust is really built on deceit is especially upsetting when it occurs in the context of a dating scam. In addition, the possibility that the event would become public, exposing the victim to scorn or sympathy, can generate profound emotions of shame. In the aftermath of a such crime, victims may be hesitant to confide in others who may otherwise provide practical and psychological help. For fear of being ridiculed or believing the police would do nothing, victims may not report such crimes. Interpersonal cyber-crimes constitute a breach of trust, and the emotional repercussions of "virtual betrayal" may be as devastating as those of physical betrayal. Victims have expressed feelings of sadness, anxiety, powerlessness, and rage. They may become despondent, even suicidal, and lose faith in other people [7, 9].

Green, Streeter and Pomeroy [10] reveal that the emotional effects of a crime and the selected coping technique rely on how well the chosen strategies meet the situational needs. "For example, if the situation resulting from a crime is perceived by the victim as somewhat controllable, he or she would be more apt to have positive emotional outcomes from using a problem-focused coping strategy as opposed to an avoidance-oriented strategy" [10, 11]. Holohan and Moos [11] further reveal that as the intensity of a stressful event grows, so does the significance of coping mechanisms. Against this backdrop, this chapter set out to foster coping mechanism, i.e. shock absorption mechanisms for dealing with cybercrime trauma while awaiting the protracted orthodox criminal justice outcomes in Nigerian setting, and by extension, other climes similar to Nigeria across the globe. The following research questions were put forward:

i. What is the classification of cybercrime?

ii. What are the patterns of cyber-crime?

iii. What are the shock absorption mechanisms adopted by victims?

2. Conceptualization of cyber crime

The concept of cybercrime is vast and the high-tech nature of the field has made it difficult for scholars in various field of cyber-criminology to agree on a particular definition for the concept. It has been suggested that since cybercrime may entail various types of crime, a definition of cybercrime has to place an emphasis on the specificity, the expertise, or the use of computer technology [12, 13]. Accordingly, Ajayi [14] observed that the above situation has made it difficult for scholars to come up with a universally accepted and recognized definition of cybercrime. Although, the definition of cybercrime varies slightly from one individual to another, there is a consensus among scholars on the important role of networked technologies in enabling this type of criminal activity. One frequently adopted definition of cybercrime describes it as any action in which computers or networks are a tool, a target or a place of criminal attack [15, 16]. The International Telecommunication Union (ITU) [17] defined "cybercrime as a criminal offence involving a computer as the object of the crime (hacking, phishing, spamming), or as the tool used to commit a material component of the offence (child pornography, hate crimes, computer fraud)". As the name suggests, this phrase appears often when computers or associated technology is employed in a criminal offence. Traditional crimes like distributing child pornography, illegal substances, and hate crimes may be considered cybercrimes since they can be perpetrated via the internet [12, 15].

Innovative Dynamic Networks (IND) [18] defined "cybercrime in a contracted sense to imply unlawful acts directed by means of electronic operations that targets the security of computer systems and the data processed by them". Cybercrime in a broader sense according to Innovative Dynamic Networks (IND) [18] refers to "prohibited activities committed by means of, or in relation to, a computer system or network, including such crimes as illegal possession and offering or distributing information by means of a computer system or network". United Nations Office on Drugs Control (UNODC), [19] defined "cybercrime as a cluster of unlawful behaviour such as offences against the confidentiality, integrity and availability of computer data and systems, computer-related offences, content-related offences and offences related to infringements of copyright and other related rights".

Cybercrime differs, according to McConnell International in Tamarkin [12], from most conventional deviant behaviours in four ways: the criminal act is relatively easy to learn, the cost of committing the crime is never proportional to the damages caused, it is a borderless crime and fourthly, the act, most times are not clearly prohibited. It is important to note that e-crime is perpetuated in the virtual environment. The virtual environment is designed in such a way that data about individuals, things, realities, proceedings, phenomena or events are depicted in mathematical symbol or any other way and transmitted through local and global networks. From the foregoing, the term 'cybercrime' can be applied in explaining, as well as, describing widespread destructions on computer data or networks through interception, interference or damage of such data or systems. It can also be used to explain and describe crime committed against computer systems or the use of the computer in committing crimes. The definitions, although not completely definitive, and perfect, provide a recognised and good framework for explaining cybercrime at the international level and within the context of this chapter.

3. Methods

This study is a narrative research that relied heavily on secondary research methodologies and examined only articles written in English. The data for this study came from a variety of sources, including books, journal essays, articles, tabloids, magazines, and scholarly materials from the internet in the areas of cybercrime. Google scholar and the Google search engine were used in the online search. This study therefore summarizes the data collected from these current research publications. After extensive research and evaluation of the available materials online, the major themes reported in this study were identified. The main ideas were selected and reported based on the frequency of their reporting in the literature. The literature focus was on reports and research findings on the curbing strategies of victims of cybercrime.

4. Classification of cybercrime

Ayofe and Irwin [20], and Poonia [21] broadly classified cybercrime into four primary taxonomics: "crime against persons, cybercrimes against property, cybercrime against organisations and cybercrimes against society".

4.1 Cybercrime against persons

In the cyber world, crime against persons manifest in form of transmitting child pornography, cyberbullying and harassment of cyber users [22].

4.2 Cybercrime against property

The second type of cybercrime includes offences committed against any and all kinds of property. "These crimes include unauthorized computer trespassing through cyberspace, Salami Attacks, computer vandalism, intellectual property crimes, transmission of harmful programs and unauthorized possession of computerized information".

4.3 Cybercrime against organisations

Organisations are seriously threatened and attacked by the activities of cybercriminals. Such attacks and threats have resulted to loss of sensitive information, money, and intellectual property. The activities of these criminals have also made consumers to lose confidence and trust in the services provided by these organisations. Cyber-terrorism is one of the most remarkable forms of cybercrime against organisations. Another form of cybercrime against organisations is cyber-warfare. Simply put, it refers to politically motivated hacking to conduct sabotage and espionage, especially, among nations.

4.4 Cybercrime against society

Cybercrimes against society may take the form of "forgery, cyber-terrorism, web jacking, polluting the youths through indecent programming, financial crimes, sale of illegal articles, net extortion, cyber-contraband, data diddling, logicbombs, etc". This type of crime also includes; revenue stamps, forgery of currency notes, certificates, mark sheets, among others, high grade scanners and printers may be used to forge these documents. Web jacking hackers obtain access to and control over the website of others, and they may even modify the content of the website to accomplish political aims or to make financial benefit.

5. Patterns of cybercrime in Nigeria

Cybercrime entails the application and manipulation of the internet to fraudulently derive benefits from unsuspecting users. Some of these crimes includes, spoofing/phishing, spamming or escrow services, web jacking and scam messages. Different authors have varying views but in simple terms, cybercrime encompasses all illegal activities carried out by a single or more individuals most times referred to as scammers, hackers, fraudsters, "419ners", using the internet through the medium of networked computers, telephones and other ICT equipment. Thus, the acts of cybercrime originated from the emergence of computers, telephones and other ICT inventions. Numerous conventional crimes are being perpetrated with the use of ICT inventions, they include:

Auction fraud: This is the misrepresentation of a product advertised for sale through an internet auction site, or the non-delivery of the products purchased through an internet auction site. The seller posts the auction as if he resides in the United States, then responds to victims with a congratulatory email stating he is outside the United States for business reasons, family emergency etc. They often post the auction under one name, and ask for the funds to be transferred to another individual or directly to him via Western Union, Money Gram or bank to bank wire transfer.

Huckstering: This is a process of obtaining email address from the internet access point using email harvesting software called web spiders (such as email Extractor Lite 1.4) to send a large number of messages to each harvested spam-trapped addresses and typical product based Spam (i.e. Spam selling an actual product to be shipped or downloaded even if the product itself is fraudulent).

Piracy: This is the act of illegally making access to people's soft copies such as, books, games, movies and CDs or DVDs, etc and make copies of same to disseminate for some gains which is usually financial gains [23]. Example, the use of pirated Microsoft Windows to install newly acquired computers; pirated home movies; and pirated MP3 music installed in phones, Ipads and other gadgets.

Hacking: This is the act of cracking firewalls or security codes with the use of computers, laptops and sophisticated phones in order to gain access to people bank accounts, data or any other profitable information.

Phishing/spoofing: This is the act of faking or forging digital information or documents [24]. Spoofing/phishing specifically connotes the fraudulent acts of forging a website to make it look like the original one to deceive persons having legitimate transactions with such websites or the harvesting of people's e-mails, and after consuming their contents, use same to defraud other people, who somehow feels the information received is authentic.

Ponzi/pyramid: This is a kind of money doubling scam. It is usually initiated as an investment for never to be receive profits. Because it a bogus and attractive investment proposal, desperate individuals often fall victim. The victims of these scams neither receives dividends nor their initial capital. Example, the 2009 money doubling scam in Calabar, Ikom and Ogoja in Cross River State.

Nigeria letter or "419": This named after a section in the Nigerian Criminal Code. 419 combines impersonation, obtaining by false pretence or advance fee fraud. The major trick for this scam is calculated persuasions. Victims are usually hooked with sensible persuasions after the fraudsters have anticipated their thoughts for every step to be taken. Victims who are charmed by these well added lies end up losing huge sums of cash or divulging their credits cards numbers or Automated Teller Machine pins.

Credit card fraud: This involves illegal or unauthorised use of people's credit/debit cards to steal their money. Out of carelessness or negligence, victims usually compromise their credit/debit cards numbers to fraudsters, who actually get same from close observation or outright theft, sometimes on gun point. In Nigeria, such numbers are obtained in ATM withdrawal terminals or robbery at any location and pins are obtained on gun point.

Identity theft: This is the act of impersonation for the purpose of committing theft. Individuals and organizational fake identities are used by fraudsters to dupe persons operating legitimate businesses.

Data dadding: This is a kind of attack which involves altering raw data just before it is processed by a computer and then changing it back after the processing is completed.

Salami attacks: These attacks are prevalent in the financial institutions. It is an alteration that is so insignificantly made such that in a single case it would go normally unnoticed. For instance, Salami Attack occurs where a bank employee inserts a program such as 'logic bomb' into the bank servers, that deducts small amount of money from the account of every customer and deposits it in another account opened and owned by the staff but with a different account name.

Internet time thefts: This is the act of manipulating or circumventing servers of network service providers in order to hack their passwords and gain login access. Fraudsters usually steal airtime from Internet Service Providers or GSM service providers, like the case that affected MTN Nigeria in February 2009.

Web jacking: This is the process of fraudulently gaining access into individual's, corporate organisation or government websites or e-mails and completely taking charge or control of it. This is done by breaking through the passwords and other unique features, and editing same whereby the original owner may not be able to gain access again.

Phone phishing: Phishing attack also extends to phones such that messages claimed to come from a bank may tell users to dial a phone number regarding problems with their bank accounts. Once the number (owned by the phisher, and provided by a voice over IP Service) has been dialled, prompts would tell users to enter their account numbers and Personal Identification Number (PIN) which the Phisher would use in defrauding the victim.

Pornography: This is the act of using ICT gadgets to illegally publish pornographic images, or the use of internet to download pornographic contents illegally. Most people still have the understanding that pornography does not amount to crime, but they forget to know that most pornographic images are illegally produced, and so everybody benefitting from such content is a criminal, while those less than 18years are delinquents.

Sale of illegal articles: This is the process of selling contrabands or illegal products such as hard drugs or weapons of mass destruction through the use of internet websites, e-mails, short message services and other means of digital communication. .

Employment/business opportunity fraud: On the Internet different websites and most often 'pop-ups' on web pages have been design to advertise lucrative employment opportunities and businesses with the aim of defrauding unemployed persons.

Forgery: This is the act of counterfeiting an original document, made possible and easy by the emergence of information and communication technology. It is the act of faking an original money note or coin, or any other document to make it look similar or almost the same as the legal or original one. In Nigeria, there are a lot of forged certificates and naira notes.

Cyber defamation: Character defamation as crime has also migrated from the verbal physical world into the digital world. This is an act of making mails, faxes, text messages, etc, slaying somebody's character and distributing same to acquaintances of the target victim. This is usually done for some selfish and fringe benefits.

Cyber theft: Any form of criminal activity that involves the use of information and communication technology is called cyber theft. Cyber theft is synonymous to cybercrime except that it is narrowed to issues that are outright theft such as, embezzlement with the use of ICT, stealing people's passwords and pins or hacking.

Cyber laundering: This is the process of illegally transferring monies or currencies with the use of ICT. This is usually done by fraudsters with the intention of concealing the source and destination of this monies or currencies. It is a common form of cyber criminality in Nigeria, where corrupt government or public officials cyber launder stolen money to other parts of the world in order to avert the wrath of EFCC and ICPC.

Key logging: This is the act of using software called key logger to stroke someone's computer keyboard in order to digitally monitor the activities taking place on that computer, and capitalizing on such information to defraud the computer owner.

Spam message: This is fake messages or e-mails directed at harvested e-mails or random numbers with persuading and attractive contents aimed at defrauding unsuspecting victims.

Malware: Malware short for malicious software, (sometimes referred to as pest ware) is a software designed to secretly access a computer system without the consent of the owner.

Lotteries fraud: Some corporate organizations in Nigeria garner millions of viewers to send text messages or to call in answers to a displayed question on TV or answer a question through SMS. The amount charged for such calls or text message (SMS) from the viewers amount to millions of Naira while the amount to be won might just be N10,000.00.

Phreaking: This type of crime involves the theft of telecommunication services, instances of which have involved using cereal box toy whistles to imitate telephone call signals, and more recently cloning mobile SIM cards. Many homes in Nigeria are now connected to DSTV products through a self made cord, which when attached to a video player, has the capability of connecting to several statements in DSTV television.

6. Victim susceptibility

Cyber criminals definitely have targets to explore. Likely cybercrime victims include;

The naive/gullible: There are certain persons who are easily deceived. This is because they trust almost everybody who comes their way. Even in the cyberspace, they try to be helpful to those they have never met. Cyber criminals prey on such individuals who are naive and slow about knowing much of the internet gimmicks. Older people usually fall into this category because they trust easily, and the intricacies of the computer are not of their generation.

Desperados (for money or "items"): So many youths wants to build sky scrapers within a twinkle of an eye and through any means, they easily fall victims to online pop-ups that read "Get rich fast". This desperation makes fraudsters to have their way with them. In many instances, these youths are cajoled into bogus and attractive business proposals, and even life-changing advertisements. In these businesses, investors who are invariably the victims never recover their initial capital, let alone make profits. The sole benefactors are the initiators of the businesses who are the perpetrators. On the other hand, students are the major victims of cyber stalkers because of their desperation to meet people online and make friendship. Sometimes, this online friendship is sought by students to boost their self-esteem or personal ego.

Unfortunately, they become victims of internet hoodlums who either manipulate them for the satisfaction of their sexual appetite or other rituals.

The inexperienced: So many people have frivolous attitude towards ICT. They are mostly contented with just making calls and sending text messages or checking e-mails, without the desire for requisite knowledge on the detail use or application of digital devices. Such persons cannot protect their phones or computers from malicious damage and intrusion. Even when they have information that such crimes exist, they still fall victims because of their inexperience.

Unlucky people: These are people who are very unlucky in life. They fall victim of cybercrimes as a result of their fate which has made them to be found at a wrong place, at a wrong time. For example, malicious viruses can be circulated in the cyberspace and only unlucky people's computers and phones will be infected, which will result to serious damage and destruction of their systems and data. In this case, it does not matter how knowledgeable or proficient you are in protecting your data, you can just be unlucky.

7. Target hardening

Cyber criminality is a phenomenon tied to the daily routine of individuals. To this end, Cohen and Felson in 1970 articulated the Routine activities approach which derives from the fact that elements of a criminal or deviant act come together in normal, legal, and routine activities [25]. At the heart of routine activities are three premises often referred to as the crime triangle; a likely offender, a suitable target, and the absence of a capable guardian. Routine activities theory posits that criminal victimization increases when motivated offenders and suitable targets converge without let or hindrance [26].

Within the cyber space, the massive spread of global system of network provides the fertile ground for the absence of a capable guardian, the internet user then becomes the suitable target, waiting to be devoured by scammers who are the motivated offenders in this regard. However, from the routine activities approach, if internet users adopt target hardening strategies like; two-step security code authentication, periodic password change, firewall settings, anti-virus definition update, One Time Password (OTP) validation, etc., online criminal victimization will decline.

Thus, target hardening ensures that online users take the responsibility to police their own activities on the internet. The stipulations of the routine activities approach is best suited to guard internet users in the sense that people are supposed to guard their login details securely from un-trusted sources. This includes ensuring that websites being accessed are well secured with the inscription "https" or "locked padlock". In another light, parental control should be activated in computer and internet gadgets used by underage children to guard them from cyber stalking and pornography. Put succinctly, the watchword to target hardening as posited by the routine activities approach with regards to cyber criminality refers to all deliberate authentication and security efforts adopted by the internet user to ensure that online activities are protected from scammers.

8. Shock absorption mechanisms

Cyber crime triggers emotional bankruptcy on victims. Most time, victims of cyber crime commits suicide upon the reality of the extent of loss they suffered in the hands of scammers. Therefore, most victims of cyber crime could have lost their sanity or life moments after reality of being duped. Shock has been found to have helped individuals regulate their emotional response mishap [27]. The aim of this chapter is to proffer possible shock absorption mechanisms which could help victims to go past the emotional scam associated with cyber crime, owing to the fact that only the living, or the sound mind can logically seek for justice and retribution for a crime committed against them. Thus, the following shock absorption mechanisms are posited:

a. Talk to someone about the situation, a problem shared is a problem half solved. There is tendency for a victim of cybercrime to reach out to someone immediately; this is to avoid being blamed. But rather than bottling up the emotion, it is better to let it out and lessen the burden in the heart.

b. Spend quality time with significant others. This will enable the victim to feel the love and moral support from close associates, than wallow in loneliness and isolation which can breed unpleasant stimuli.

c. Approach daily tasks with care. In the midst of clouded thoughts, accidents are more likely to happen after severe stress. Thus, the need to undertake activities with care. It is most appropriate to have adequate rest for the day, until the victim gradually bounce back to normalcy.

d. It is necessary to re-establish a normal routine as soon as possible, but it should be taken gradually.

e. Exercise should be undertaken at certain intervals of the day, as this could be shuttled between periods of relaxation

f. Victims are not to take laws into their hands but rather should report to law enforcement agents to take action.

The above shock absorption mechanism require that victim avoid certain behavioural tendencies such as: alcohol or drugs for the purpose of relieve from emotional pain, making substantive life decisions at the moment, withholding emotions and self-blaming. It is pertinent to note at this juncture that cybercrime could leave victims with a feeling of emptiness, and a trigger of emotion even after several years. However, in this midst of these, it is impossible to undo what has happened but life can be good again in time. In the light of the forgoing, significant others to the victim have traditional roles to play to promote the integral efforts of shock absorption. This includes spending time with the victim, offering assistance where necessary, giving listening ear, avoid triggering negative emotions on the victim, showing empathy, and any other humanistic gesture.

9. Conclusion

Cybercrime since its inception has left its victims shattered and demoralized to the point of taking their own life or loosing total sanity to the point of no recovery, in a word, cyber crime has left its victims in a state of Robert Merton's "anomie". The peculiarities of cybercrime lie in the fact that the victim willingly lands him or herself into it without being forced to do so. It starts with what seem to be a friendly conversation and exchange of correspondences and pleasantries which turn into a scamming spree. Unlike other criminal ventures, cyber criminality stem from betray of "trust" but unfortunately "false trust". To this end, victims are left battered and shattered, and could act irrationally against own-self before state actors set out to track the offender. For this, this chapter outlined shock absorption mechanisms to deal with the rising and dynamic trend of cyber criminality to save the victim prior to state intervention to bring the perpetrator to book.

Acknowledgements

The authors appreciate Ifeoma Nuela Arinze who typeset the initial draft of the paper.

Conflict of interests

The authors declare no conflict of interest.

Notes/thanks/other declarations

The authors sincerely appreciate sources cited in this intellectual output.

Author details

Obinna J. Eze[1], John Thompson Okpa[2], Chukwuemeka Dominic Onyejegbu[3] and Benjamin Okorie Ajah[1]*

1 Department of Sociology and Anthropology, University of Nigeria, Nsukka, Nigeria

2 Department of Sociology, University of Calabar, Calabar, Nigeria

3 Social Sciences Unit, School of General Studies, University of Nigeria, Enugu, Nigeria

*Address all correspondence to: okorie.ajah@unn.edu.ng

References

[1] Alawari BM, Ajah OB. Understanding the Gender Dimensions of Cyberbullying among Undergraduates in Nigeria. Zaria: Ahmadu Bello University Press Limited; 2017

[2] Reep-vanden Bergh CMM, Junger M. Victims of cybercrime in Europe: A review of victim surveys. Crime Science. 2018;**7**(1):5

[3] Okpa JT, Ilupeju AA, Eshiotse E. Cybercrime and socio-economic development of corporate Organisations in Cross River State, Nigeria. Asian Journal of Scientific Research. 2020;**13**:205-213

[4] Lenhart A. Teens, Social Media, and Technology Overview. Washington, DC, USA: Pew Research Center; 2015

[5] Okpa JT, Ajah BO, Nzeakor OF, Eshiotse E, Abang TA. Business e-mail compromise scam, cyber victimisation and economic sustainability of corporate organisations in Nigeria. Security Journal. 2022;**35**(2):1-23. DOI: 10.1057/s41284-022-00342-5

[6] Ukwayi JK, Okpa JT. Critical assessment of Nigeria Criminal Justice System and the Perennial Problem of awaiting trial in Port Harcourt Maximum prison, Rivers State. Global Journal of Social Sciences. 2017;**16**:17-25

[7] Reyns BW, Fisher BS, Bossler AM, Holt TJ. Opportunity and self-control: Do they predict multiple forms of online victimization? American Journal of Criminal Justice. 2019;**44**(1):63-82

[8] Okpa JT, Ajah BO, Igbe JE. Rising trend of phishing attacks on corporate organisations in Cross River State. Nigeria International Journal of Cyber Criminology. 2021;**14**:460-478

[9] Ajah BO, Onyejegbu DC. Neo-economy and militating effects of Africa's profile on cybercrime. International Journal of Cyber Criminology. 2019;**13**(2):326-342

[10] Green DL, Streeter C, Pomeroy E. A multivariate model of the stress and coping process. Stress, Trauma and Crisis. 2005;**8**(1):61-73

[11] Holohan C, Moss R. Life stressors, personal and social resources and depression: A four year structural model. Journal of Abnormal Psychology. 1990;**11**(1):31-38

[12] Tamarkin E. Cybercrime: A complex problem requiring a multi-faceted response. ISS Policy Brief. 2014;**51**:1-3

[13] Nnam MU, Ajah BO, Arua CC, Okechukwu G, Okorie CO. The war must be sustained: An integrated theoretical perspective of the Cyberspace-Boko Haram Terrorism Nexus in Nigeria. International Journal of Cyber Criminology. 2019;**13**(2):379-395

[14] Ajayi EFG. Challenges to enforcement of cyber-crimes laws and policy. Journal of Internet and Information Systems. 2016;**6**(1):1-12

[15] Ndubueze PN, Igbo EUM. Third parties and cyber-crime policing in Nigeria: Some reflections. Oxford University Press. 2013;**8**(1):59-68

[16] Cisar P, Maravic CS, Bosnjak S. Cybercrime and Digital Forensics–Technologies and Approaches. Vienna, Austria: Daaam International Scientific Book; 2014. pp. 525-542

[17] International Telecommunication Union (ITU). Understanding

Cybercrime: A Guide for Developing Countries. Switzerland: ITU Publication; 2009

[18] Innovative Dynamic Networks (IND). United Nations' definition of cybercrime. 2016. Available from: https://idn-wi.com/united-nations-definition-cybercrime/

[19] United Nations Office on Drugs and Crime (UNODC). Cybercrime. 2018. Available from: https://www.unodc.org/unodc/en/cybercrime/global-programmecybercrime.html

[20] Ayofe AN, Irwin B. Cyber security: Challenges and the way forward. GESJ: Computer Science and Telecommunications. 2010;**6**(29):56-69

[21] Poonia AS. Cyber crime: Challenges and its classification. International Journal of Emerging Trends & Technology in Computer Science (IJETTCS). 2014;**3**(6):119-121

[22] Desai PN, Patel AM. Cyber Crime against Person. International Journal of Innovations in Engineering and Technology (IJIET). 2013;**2**(3):198-201

[23] Longe OB, Longe FA. The Nigerian web content: Combating the pornographic malaise using web filters. Journal of Information Technology Impact. 2005;**5**(2):29-50

[24] Loftness S. Responding to "phishing" Attacks. USA: Colenbrook Parnera Publishers; 2004

[25] Schaefer S. The Victim and His Criminal: A Study of Functional Responsibility. New York: Random House; 2005

[26] Brown SE, Esbensen F, Geis G. Criminology: Explaining Crime and Its Content. USA: Mathew Bender & Company Inc; 2010

[27] Pearlin LI, Schooler C. The structure of coping. Journal of Health and Social Behaviour. 1978;**19**(March):2-21

Chapter 2

Classification Models for Preventing Juvenile Crimes Committed with Malware Apps

Joshua Ojo Nehinbe

Abstract

Spectacular developments that were recorded in the field of software engineering in recent years have led to the influx of software industry with series of computer apps such as dating apps, games apps, entertainment apps, banking apps, Photoshop apps, meetings and virtual conferencing apps. Studies have shown that most computer apps are widely accessible to adults and juveniles to download and effortlessly navigate through them. However, researchers have now revealed the existence of malware apps as new groups of computer apps that are strongly competing with legitimate computer apps and the latest rates at which some juveniles can adopt them to commit crimes. These discoveries have raised serious doubts about the elements of the crimes, the circumstances that surround vulnerable children to commit the crimes and how these dilemmas are rarely buttressed by pragmatic studies over the years. This chapter adopts mixed methods to critically explore the above issues. Qualitative interviews of 60 teenagers (between the ages of 10 and 17) and 20 grown-up children (between the ages of 18 and 22) together with 5 professionals were carried out. The analysis extended the generic elements of juvenile crime and raised new legal dilemmas regarding the concepts of transfer of criminal liability, compelled (or obligated) liability, 'act' that constitutes juvenile crimes and the restrictive applicability regarding criminal consent of extremely young children that are still under the tutelage and guidance of their parents.

Keywords: crime, juvenile crime, Malware apps, intruders, computer crime, suspects

1. Introduction

The spectacular developments recorded in the field of software engineering over the years have directly led to the influx of the software industry with series of computer apps such as dating apps, loan apps, games apps, entertainment (or music) apps, image editing apps, cloud data storage apps, messaging (or messenger) apps, discord apps, virtual workspace apps; password managing apps, scanning apps, antivirus apps, backup apps, video streaming apps, TV shows apps, codes and scripting apps, language learning apps, banking apps, cloud apps, presentation apps, reminders apps; apps for apps designers, Photoshop apps, meetings (or virtual) conferencing apps, zip file apps and task-management apps [1, 2]. The beauty of these computer apps is that significant numbers of them are widely available and accessible to adults and juveniles to download and easily navigate through them. Nonetheless, market research has shown the existence of malware apps that are identified as new groups of computer apps that are strongly competing with legitimate computer apps in the software industry in recent time [3–5].

Serious dilemma has also been expressed in recent years on the fact that the scales of the operations and the levels of data security that companies and their respective service providers can jointly offer to customers (users of their services) across the globe are not the same. These tropical issues have pointed out the likelihood of future debates on the usage of criminal laws and or civic laws to adjudge corporate liability and the impact of overseas jurisdictions in criminal proceedings regarding juvenile committed crimes with offshore (or onshore) malware apps. Besides, juvenile crimes are long-standing social problems but the recent surveys have further compounded the problems [3, 4, 6]. Several empirical conclusions have now shown that the above advancements are facing serious threats and firm criticisms across the globe given the prevalent cases and sudden surge in the menace at which some juveniles can now adopt malware apps to commit crimes [7, 8].

Fundamentally, contemporary studies believe that juvenile is a social concept to describe adolescent or the characteristics of children and young people that have not fully attained the age of maturity [9, 10]. Children that are below the age of 18 are mostly categorized as juveniles in most societies [11]. A crime is defined as an act that is prescribed to be unlawful by statutes or criminal laws of a state or country [12–14]. Therefore, juvenile crimes connote unlawful acts that are committed by the children below the age of 18 years in the society [15, 16]. Malware apps are malicious software apps that are purposely created to have intrusive or criminally-minded motives. Malware apps, malicious software apps and malicious apps will be interchangeably used to connote the same concept in this chapter. Studies have further raised classical deliberations concerning the underlying motives of the designers of malware apps and their users. Studies on the capabilities of malware apps have shown that they can corrupt, explore, delete, deface, swindle and steal resources or personal details of another people. Some malware apps intend to distort the clarity of pictures and render them blurred. Empirical studies on software engineering have shown that malware apps can mute telephone conversations and suddenly turn them to be speechless or voiceless. It is a well-established fact that malicious software apps can make audio conversations inaudible to the participants in virtual conferences, virtual interview and virtual lectures. Another classical issue here is that the victims of various crimes that some juvenile can commit with malware apps may impact on just a person, two persons, more than two persons and more than one governmental body.

Several topical studies in the domains of criminal laws have also substantiated and categorically stated that juvenile crimes are social problems [10, 15]. Tropical issues in criminal laws have led to broad explanations of various elements of crimes and several considerations required charging or acquainting the suspects that are alleged of crimes over the years. However, apart from the fact that the concept of juvenile crimes committed with malware apps are emerging issues in the global society, there are clearly invalidated studies that have substantiated the correlations between the elements of these kind of crimes and the circumstances that surround the vulnerable children that may be alleged of the crimes, even in recent years [17]. The problems with these challenges are enormous. For instance, the actual causes and determinants of the above crimes are prone to oversight. The fear is that failure of some parents, guardians and regulatory bodies to notice and counter the waves of the juvenile crime at an early stage may suddenly aggravate the effective supervision and 'the perceived level of misunderstanding of various forms' of the act [8]. Notwithstanding, social orientation, parenting, policing and social policy can begin to face strict criticisms in most civic settings if more and more youths are alleged of the juvenile crimes committed with malware apps [7, 8]. Agitations that will be calling for serious awareness through media and political participations may begin to surge in order to wakeup parents, guardians and police in their duties so that they can perfectly monitor vulnerable juveniles that may adopt malware apps to perpetrate crime or be victims of the crimes. The severity of the above problems is worrisome due to the inabilities of most governments and agencies to design suitable preventive interventions that will decisively curtail the emerging developments in the software industry. For these reasons, parenting and parental care as well as the efficacies of several social policies across the globe is generating increasing criticisms [8].

Another critical consideration is that most of the users of modern apps (especially employees) have little or no choice to determine the level of their participation and the information that they must supply to enable them use some of the modern apps that currently exist in the software industry. The reason is that most employers rarely involve their employees in the selection and evaluation of modern apps for supporting process flow at workplace. Instead, some employers often obliged their employees to use and wholeheartedly adopt them especially for scheduling, management of task, workplace meetings, virtual events (e.g. conferences, workshops and lectures), official reporting and chatting with colleagues or bosses, etc. with the view of boosting productivity, sales and profitability. Given the fact that Internet and all computer apps have inherent vulnerabilities, bugs and vulnerabilities that may not be conspicuously known to software experts to fix, on these reasons, the security of the entire components of the above apps and the level of the protection they can offer to the end-users that have trusted (or were compelled to trust) and sincerely adopted them begin to attract the attentions of criminologists, legal and security experts [8, 18].

Industrial dispute and conflict resolution among software companies, victims and service providers and how to amicably resolve them are now raising serious legal and technical debates especially, if it appears that computer apps that users sincerely trusted have aided or assisted criminals to achieve their malicious motives. The above legal and software issues are raising legal dilemma about the issues of compelled (or obligated) liability due to an assurance from a third party applications and the actual organization that an employee that incurs harm from computer apps can actually sue between their employer and the vendors of computer apps that seems to have aided the crime. Some school of thought may believe that service providers can be penalized by law for not doing their jobs well. The fact is that employees may be punished for refusing to embrace the modern apps obliged by his/her employer. So, another school of thought may argue the above scenario on the basis of the 'commission' or 'omission' as a possible element of the crime. In this circumstance, the question of whether the law is justified to penalize or absolve the employer and or their service providers, and then dismiss or award compensation(s) to the employee requires deep legal technicality of these new kinds of cybercrimes [12–14].

Despite of the existing standards of software methodologies, it is impossible to rigorously protect different computer apps in the same way and especially given the fact that computer apps are designed and manufactured by different vendors [19]. For instance, copyright, Intellectual Property (IP) and trade laws usually prohibit two different software companies from copying their designs, underlying theories, principle and cryptographic algorithms that they adopt to protect the passwords and messages in transit. Thus, some computer apps will surely have inbuilt security features than another [19]. So, the level of computer crimes that intruders can commit with them equally varies. The fear is that collaborative empirical reviews had earlier warned that the above problems can increasingly metamorphosed into complex problems that will compound the detective and preventive interventions for various kinds of juvenile crimes committed with malware apps in the next decade if the global society erroneously allows them to escalate to high levels [8, 17]. Consequently, the initial motives of the designers of computer apps have begin to suffer wider criticisms in a recent time.

Another puzzling dilemma that can be confronting most software and legal experts is to establish the correlations between the elements of the above juvenile crimes with malware apps and the circumstances that surround the vulnerable children that are rarely buttressed and made explicit by empirical studies on juvenile crimes over the year [17]. For this reason, the objectives of this chapter are split into three groups. The chapter intends to explore the causes of juvenile crime committed with malware apps. The chapter also intends to state various 'act' that can constitute infringement and classified as juvenile crimes committed with malware apps. The chapter intends to adopt the above objectives to propose empirically proven classification models to simplify and explicate the above legal and social dilemmas.

By using mixed methods and quantitative interviews, 60 teenagers (between the ages of 10 and 17) and 20 grown-up children (between the ages of 18 and 22) together with 5 professionals were recruited to explore the above social problems. Quantitative analysis of logs of Snort Intrusion Detection Systems (SIDS) was incorporated into the mainstream of the sessions of the interactions with the above participants. Thereafter, we thematically analyzed the results obtained. One of the contributions of this chapter is that it has extended the generic elements of juvenile crimes and further suggested various determinants of the circumstances that may surround vulnerable children to commit juvenile crimes with malware apps. The chapter has further empirically substantiated the correlations between the determinants of the circumstance that may surround vulnerable children and the elements of juvenile crimes with malware apps. The paper also introduced new legal discourses and then offered suggestions to lessen parenting hurdles and how to countering the weaknesses that may be inherent in the policing and social policies on the above category of global crimes. The remainders of this chapter are organized as follows. Section 2 discusses the domain of coverage of modern computer apps. The section also opens up new legal dilemmas on computer apps. Section 3 explains the rudiments of juvenile crimes committed with malware apps. The section further itemizes the generic elements of most crime scenes. Section 4 discusses the methodology of the survey. Section 5 states and analyses the results and their implications. Section 6 concludes the chapter. The chapter also offers suitable areas that researchers can explore to extend the research that is reported in this chapter.

2. Modern computer apps

Computer application programs are often abbreviated as computer apps [1]. Legitimate computer apps in the software industry can serve a wide collection of functions and purposes to the target audience [1, 2]. Some legitimate apps enable their users to view, upload and share favorite songs and new albums of artists with friends and social groups. There are modern apps that enable people to have access to audio-books, historic podcasts, video, motion pictures, selected artifacts and read, examine or watch them at their leisure time. Some modern computer apps enable users to create their personal accounts, sign on to the apps by using their personalized accounts and remotely connect to some companies, colleagues and professionals in other locations and engage in virtual conferencing.

The functionalities of some of the accessible modern apps in the software industry allow their users to also add or invite people (users) as contacts [1]. Some computer apps allow users to select the kinds of services (such as message or call) of their choice. Some apps can enable users to equally share files and log on to chat box and engage in private or official conversations. Some of the existing apps enable their users to customize and fine-tune their background information. Some computer apps permit users to upload their personal pictures and insert personal notes or personal ideologies on their chat profiles. Some modern apps can engage participants and host many interactive sessions. Besides, some computer apps can engage several participants in every session and they will still experience clear pictures and audible audio conversations.

More so, the current advancements in computer apps cut-across scores of human domains. For example, there are computer apps for modern photography subsumes taking pictures (photographs or photos, snapshots), cinematography (movies production), film production, picture production, animatronics (animation), computer graphics, shooting cartoons, moving picture, mood mining, printing and camera repair. Furthermore, legitimate apps such as printing apps can enable both the professional and amateur printers to increase their creativities and proficiencies in photography. Nevertheless, the striking issues on modern computer apps are worrisome. The issue of privacy control and virtual data sharing syndrome have made some experts to reserve their comments on the confidentiality, integrity, availability and non-repudiation of virtual signals that would have migrated through several networks in different intercontinental boundaries in the course of using most of the existing legal computer apps. Another concern is that unproven juveniles that are suspects of the intrusions into computer apps may (or may not) necessarily commit the offense. Some juveniles may not possess software engineering skills that they require to design apps that will conform to best global standards. For these reasons, the foreseeable impacts and the confidence that users usually have in computer apps are constantly threatening the trust, customers' loyalty, recommendation and continuous usage of the current groups of computer apps in the software industry.

The legal interpretation and the technicality of cases of violations and misdemeanors regarding modern computer apps have now raised four pondering issues that require lateral deliberation and special attentions [3]. Firstly, the emergence of accidental (inadvertent, unintentional or unplanned) damages that can be incurred by end-users based on "the services and the trust in computer apps", especially if the events are proven (or suspected) to be directly caused by the occurrence of unexpected intrusions against the modern computer apps they use must require in-depth legal consideration. Secondly, the criminal consent of underaged children demands urgent review in law books and contemporay bulletins. Thirdly, the issue of transfer of criminal liability in the circumstances of using an apps and getting into "avoidable trouble" and fourthly, the circumstances whereby some legitimate apps may be held (or charged) by complainants for being liable to have criminally permitted some juvenile crimes (that left the victims with severe impacts) to have permeated (infused, pervaded) or spread through them. After all, some statutory laws legalize complainants to institute legal actions against some manufacturers of items (or products) for the accidental (involuntary or unexpected) damages they have incurred by virtue of the trust they have in their products (or services) and in the course of using their products, services or items. The above legal paradigms and discourses are new debatable issues that we have put forward in this chapter to the criminologists and legal experts in the domains of computer apps to critically explore [8].

3. The fundamentals of juvenile crimes committed with malware apps

The crime itself and its elements are two inclusive components of juvenile crimes with malware apps. Unlawful acts that can constitute juvenile crimes are subsets of cybercrimes [7]. Studies on minor offenders converge and affirm that juvenile offenders are mostly tried in juvenile courts [15]. Cybercrimes are various crimes that perpetrators commit by means of digital devices (such as computers and mobile phones) and Internet facility. However, most studies on juvenile crimes that relate to cybercrimes are inexplicit and they are often reserved on the various manners that vulnerable children can be accused of breaking (or to have attempted to break) cyber laws with malware apps [11]. Rather, most contemporary studies simply treat and group the majority of the children that are alleged of cybercrimes as minor offenders (or minors that are suspects of cybercrimes).

Investigations of crime and the outcomes of the proceedings of criminal courts can be used to classify juveniles that are held as suspects of crimes committed with malware apps [10]. Basically, an offender is a term that represents a lawbreaker, delinquent or criminal. Thus, juveniles are treated as minor offenders in criminal courts [17]. A child that has been proved by courts of competent jurisdiction to have contravened cyber law(s) for the first time is known as first-time offender in cybercrime. In the same way, a child that has been proved by courts of competent jurisdiction to have consistently contravened any section of cyber laws is often called habitual offender in cybercrimes. In terms of recovery strategy, the above two groups of juvenile offenders obviously require different therapeutic interventions. In effect, rehabilitative and punitive interventions for the above settings must be commensurate to the offenses committed by offenders in order to strictly comply with Human Right laws.

3.1 What is juvenile crime committed with malware apps?

There are lots of juvenile crimes and many reasons that may lure or encourage young children to commit crimes with malware apps [10]. The bottom line of these issues is that any act of infringement by juvenile with malware apps is a crime in this respect. Infringement can be defined as an act of violation or misdemeanor and such act usually disregards ethical agreement or moral uprightness. For examples, a child that uses malware apps to unlawfully break into the telephone or computer system of another person, or steal, corrupt, modify or update the information in the electronic device(s) of another person can be alleged of crimes committed with malware apps. Additionally, a child can be alleged of juvenile crimes with malware if he/she decides to send unsolicited and offensive mail(s), insulting text(s), disgusting image(s), nasty call(s), threatening call(s), intimidating mails, etc. with malicious apps to a person that regards the 'act' as offensive and feels insulted with the 'act'. In addition, a child that uses unapproved apps to share and disseminate the picture(s) or blog(s) of another person, (either knowingly or unknowingly) in an offensive manner, or uses malware apps to steal the identity of another person with malware, or he/she uses malicious apps to sell contraband items or stalking innocent person(s) with malware can be alleged of juvenile crimes with malware apps.

Moreover, a child that uses malware apps to disrupt the business operations of a private person or corporate organization(s), or uses malware apps that behave in the manner that resembles malevolent computer programs (either local or indigenous computer program(s), foreign or proprietary program(s)), can equally be alleged of committing juvenile crimes with malware apps. Not only that, a child that uses malicious apps to harm (or intend to harm) another person(s), or uses any malware apps to spread computer viruses, Trojans, worms, etc. across computer or mobile networks may be alleged of juvenile crimes with malware apps. Fundamentally, a child that uses malware apps to unlawfully install (or he/she is caught while requesting for information to install) spyware in a computer or mobile phone of another person or organization's networks can be alleged of committing juvenile crime with malware apps. There are strong contestations pertaining to the legality of the fact that juveniles are socially categorized as adolescents, minors or teenagers in civilized societies. For this reason, contemporary studies argue that the ages of juveniles can technically qualify them to be accorded with the same treatment and honor that characterized the young people that their ages belong to the beginning of puberty and maturity age in the society. With these stacks of controversies, the statutory consent of the alleged minors or teenagers to be 'old enough' to discern (or must have known) malware apps and all actions that premeditate juvenile crimes in the process of using computer apps require rigorous legal interpretations.

3.2 Elements of juvenile crimes with malware apps

Studies have made known that juvenile crimes committed with malware apps can involve many elements [10, 20]. The term elements of juvenile crime describe various circumstances that surround or underlie the crime that are defined and recognized by the statutes or laws of a given state (or nation) or international court of justice. In other words, the statutes of each sovereign state usually set what should be the elements and the limit of juvenile crimes. However, studies have shown that juvenile crimes with malware apps are novel areas that have not been completely proved in pragmatic manners over the years.

Figure 1 illustrates the generic (standard) elements of juvenile crimes with maware apps [17, 21]. According to Britannica [22], consideration of the generic elements of juvenile crime should revolve round the conduct of the child (or criminal act), the mental state of the mind of the accused child at the time of the conduct (criminal intent), concurrence (agreement/disagreement to perform the 'act') and the causation between the conduct of the child (act) and the effect of the 'act' (criminal liability) either there is victim (s) or there is no victim involved (in case of victimless crime) [23].

Nonetheless, some of the social and legal components of the above elements of juvenile crimes remain debatable over the years [17]. The premise of this chapter is that investigators, prosecutors and judges must not expend huge resources to clearly unravel juvenile crimes committed with malware especially if the crimes are not properly split into their elements. The above standard elements are also common to

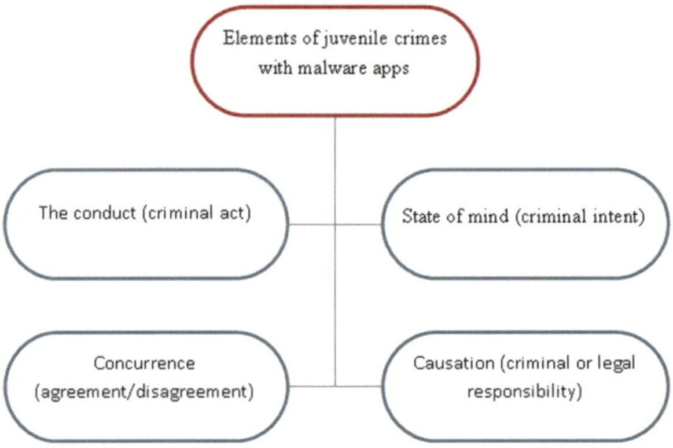

Figure 1.
Generic elements of juvenile crimes with malware apps.

most crimes. They are usually defined in criminal statutory books but prosecutors must be able to prove them beyond reasonable doubt in order to prosecute suspects of juvenile crimes. The above elements of juvenile crimes committed with malware apps may connote omission (neglect) by a child (voluntary act), commission (command0 (state of mind) of a child at the point of committing the act. A child can commit an act with malware apps voluntarily (intentionally) or involuntarily (unintentionally). For example, a child that intentionally uses malware apps to commit any of the above categories of crimes or related crime(s) is an example of voluntary intentional crime. Conversely, a child that unintentionally uses malware apps to commit any of the above categories of crimes or related crimes is an indication of involuntary or unintentional crime.

The minimum age that a child can be criminally held responsible for a crime usually varies but the internationally recognized range is between 6 and 18 years [16]. Furthermore, in a society where comprehensive justice systems exist for minor offenders (or may not exist), the prosecutor must be able to establish the legality of the fact that the child is criminally liable to be held responsible for the act especially if the allegation involves severe damage that is attributed to a child that is still below the above age range. That is, a prosecutor must be able to proof that it was through the direct participation of the child with malware apps that have done the harm or expected to have done harm to the victims(s). Detectives must be able to equally talk emphatically (while necessary) about the alleged malware apps must have been the root cause(s) of the allegation(s), or the investigative reports have traced or attributed the alleged malware apps to the accused child without any reasonable doubt.

Investigators may hold some children criminally liable (responsible) for certain juvenile crimes committed with malware apps for different reasons. Investigators may hold two or more children criminally liable for certain juvenile crimes in the above context if there are sufficient evidence, feelings and likelihood of accomplice in the allegation. Shared or joint criminal liability is a current legal issue in juvenile crimes with malware apps. This concept becomes relevant whenever a child deliberately (inadvertently) fails to act or prevent the act when in actual fact it is the legal duty of the child to act and given the fact that the child is capable of mitigating the crime at that particular instance. Ignorance is not acceptable in criminal laws. The question is whether the child is legally educated enough to ascertain and remove the uncertainties that surround the level of his/her "inadvertently act" through the unawareness or awareness of his/her legal duties in the society. The second issues is whether the child can be charged (or not charged) for his/her failure to take action (or not to take action) in the context of criminal liability" are purely determined by the prosecutor, judge and the defendant(s) of the case.

Some criminal laws may not actually forbid children from playing with the mobile phones and Internet-enabled computers of their parents, guardians and fosters at home. So, the determination of the mental element (guilty mind) of the perpetrator (or a child accused of juvenile crimes committed with malware apps) and the issue of "joint criminal liability" among two or more children require rigorous exercises. This is because studies have shown that most statutory documents have not clearly clarified the concept of "guilty mind" in the above context.

A child that was playing with his/her parent's phone may not have 'purposely', 'recklessly', 'knowingly', 'negligently', 'carelessly' or 'neglectfully' infringed the privacy of another person. The rationale for accusing, adjudging or acquitting a child of being guilty (or not guilty) of malware crimes may seem absurd in legal and media settings if the elements of the crime are improperly substantiated beyond reasonable doubt. The approach that media adopts to present and deal with the issue of juvenile crimes with malware apps and transmit them to the society worth consideration. This understanding is that the media has the power to reshape the youth and the society's knowledge and understanding about juvenile crimes and social issues that come with them. These can in turn influence the formulation of regulatory legislations and their interpretations on the above social and legal issues. In essence, juvenile crimes committed with malware apps are subsets of other kinds of global crimes committed by vulnerable young and adult people. Yet, with inadequate legal frameworks in the global community, recent topical studies have shown that there are several refutable uncertainties about the above concepts and the statistical relationships of the entities that essentially constitute and underlie the various elements of the juvenile crimes committed with malware apps across the globe [16].

4. Methodology

The researcher recruited 60 teenagers (between the ages of 10 and 17) and 25 grown-up children (between the ages of 18 and 22) for the survey. The datasets for the survey were collected with the help of mixed methods and quantitative virtual interviews using emails and Whatsapp conferencing tool. The participants were presented with two different forms of the logs of Snort Intrusion Detection System (SIDS) in four virtual conferencing sessions to evaluate their level of understanding on network forensic investigations of computer and related crimes.

The first category of the logs of SIDS was raw forensic evidence of two different trace files that were collected from the spanning mode of the computer networks of a University for a period of 120 hours. **Figure 2** illustrates the second categories of the datasets and how forensic investigators can design log analyzers with C++ programming language to investigate forensic evidence.

The quantitative analysis of the above logs were presented to the children in four brainstorming sessions and the conversations focused mainly on crime investigations and how it is also possible for detectives to easily track and arrest the suspects of crimes committed with malware apps, cybercrimes or computer and related crimes. The statistical probabilities of the themes of the responses obtained from the

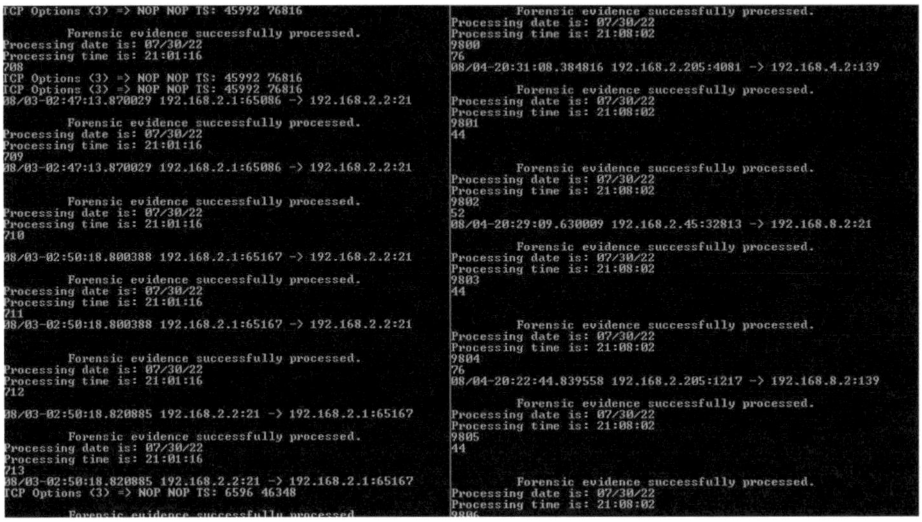

Figure 2.
Evidence of log analysis of forensic evidence that can indicate cybercrimes.

participants were also analyzed. The probability of a theme is the likelihood that the event will occur in a collection of other themes.

$$\text{Probability } (P) = \frac{\text{Number assigned to theme (event)}}{\text{Total number of themes (events)}} \tag{1}$$

In addition, the probability of occurrences is interchangeably used as the prevalence index in some instance in this chapter. The most significant of the findings in the above investigations are presented below.

5. Results and analysis of malicious apps

Virtual demonstration of crime investigation that was carried specifically discussed a total of 9805 forensic messages that may be indicative of unlawful activities in cyber laws with the participants. Further simulations showed that there are intrusion detectors, such as Snort IDS that organizations can install within the gateway to their networks in order to instantly report and log malicious events as they are occurring or migrating in and out of the networks with the participants. The experiments simulated that SIDS is capable of tracking malware apps and to further expose potential perpetrators of illegal probing of cyber physical systems. The results also showed the existence of malicious apps that intended to spy specific ports of digital devices, unlawful propagation of packets with extremely long parameters; intrusions with invalid File Transfer Protocol (FTP) commands at Disk Operating System (DOS) command line, intrusions that intend to flout the loading of certain web pages and hypertext links and intrusions that spy the Hypertext Transfer Protocol (HTTP) addresses were tracked by intrusion detectors. Furthermore, it was demonstrated that detectives can trail and arrest all the suspects of malware apps through the sources of malicious activities. The sources (or addresses) of illegal FTP messages that have packets with extended payloads that exceed beyond the maximum length of packets that have been set up to migrate within the computer networks of a university, characteristics of malicious attempts that resemble criminal attempts to overload digital networks with deformed packets and the evidence of the detection of FTP command parameters that were malformed but intruders decided to overload the networks with them were spotted by the participants in the presentations. In addition, the simulation also revealed how it is possible for SIDS to detect unlawful attempts to crack the passwords of users of computer apps (or other software) with motives to gain illegal access and steal the passwords of users in the host that is running services such as Trivial File Transfer Protocol (TFTP). In addition, it was shown that the above toolkit can equally monitor, log and report malicious activities that indicate bad sessions, computer attacks on the Server Message Block (SMB) and intrusions that aim at slowing down (or delaying) the inter-process communication between two processes that can be running as background processes in different networks.

Additionally, from the responses of the participants, we identified and classified prominent malware apps on the basis of their coaching skills. The prevalent of fake coaching apps were sought in **Figure 3**. The observations pointed out that there were no clear winners among fake career coaching apps, deceptive dating apps, fake financial coaching apps and fake voice coaching apps. We also observed that fake moral coaching apps were not frequently detected and understood by significant numbers of vulnerable children. The figure suggests that fake career coaching apps has the prevalent index of 0.235 among the list of categories of common manware apps that are competing with legitimate apps in the software industry. Fake career coaching apps pose to be the meeting point of opportunities for progressive person in life. These apps offer career counseling on prosperous occupations (or lucrative work) that a person can undertake for a significant period of time. We noted that some of the variants of these malware apps also focus on personal development and self-development in chosen occupations. Respondents believed that some of these apps assumed the roles and respondibilities of professionals with the intention to offer useful advice and information on available vocations, job vacancies and employment opportunities in certain localities.

Similarly, the above empirical survey suggests that fake moral coaching apps has the prevalent index of 0.176 among the categories of manware apps that are competing with modern apps in the software industry. Fake moral coaching apps profess to be experts in moral discussions and ethical dimensions in the civic society. These apps are designed to present principles of right and wrong behavior to the target audience. They pretend to frown at dishonest and immoral acts in the society. Some of the existing fake moral apps are holding or manifesting themselves as the authorities with high principles and

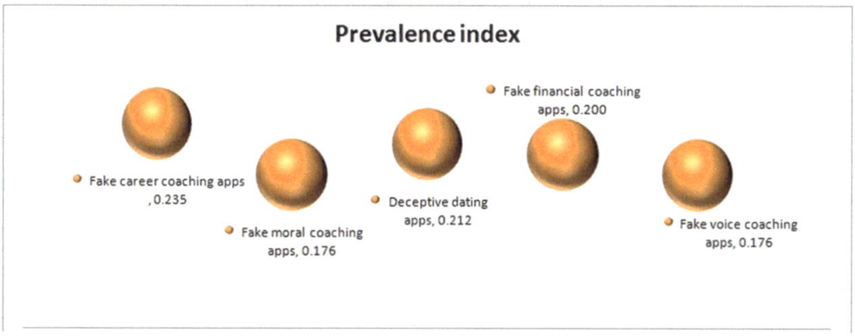

Figure 3.
Categories of malware apps competing with legitimate apps.

custodians of high moral standards that online users should consult whenever they need to seek for advice, directions and suggestions on proper conduct. The variants of these malware apps often pride themselve on being highly educated and knowledgeable in comptemporary issues, moral and ethical values whereas, they are dishonorable apps and their underlying intention is to radicalize young children. Fake moral coaching apps achieve their malicious motives by gradually focusing on discussions about radicalization like ideological concepts, religious fanatism and social cricism.

Figure 3 also put forward the idea that deceptive dating apps has the prevalent index of 0.212 among the categories of manware apps that are competing with genuine apps in the industry. Significant numbers of the respondents believed are several fake dating apps in software industry. The results showed that some fake dating apps can pretend to offer courtship, sex education and dating tips that will assist young children to find the right partners and improve their successes in dating and future relationships. These malicious apps can pretend (or operate) as if their ideas, theories and concepts will definitely assist young childre to build happy, satisfying and successful future relationships. Some fake dating apps have inbuilt stages of romantic relationships for two individuals to engage in romantic activities together. Some fake dating apps come with criteria that vulnerable youths can adopt to evaluate their partmner's suitability for future intimate relationship. The studies showed that the variations of the above apps can adopt romantic terms and slogan to trill vulnerable kids.

These above observations recommend that fake voice coaching apps has the prevalent index of 0.176 among the categories of manware apps that are competing with justifiable apps in software industry. Fake voice apps are manware apps that pretend to be coaching children on mastery of sounds, accent and mode of speech in certain settings owing to adopt the data they gather from vulnerable children to defraud some people and swindle some vulnerable friends and the relatives of the users in later time. Some of these fake coaching apps can also request patronizers to introduce the apps to their friends and relatives. But then, the apps are secretly devicing strategies to impersonate their users by using their voice, accent, pronunciation and patterns of intonation of the victims. Thereafter, the apps may masqurade and cause serious conflicts between the vulnerable children and their relatives. These apps pretend to be experts in vocal chords, pronounciation and sounds. Some of these manware apps indirectly introduce children to certain suspicious phone numbers that can require bio-data and bank details from the targets.

The observation also implies that fake financial coaching apps has the prevalent index of 0.200 among the categories of maware apps that are competing with genuine apps in the software industry. These groups of waware apps pretend to be experts that are willing to assist indigent children to overcome their financia difficulties and attain their aspirations in life. Sometimes, these apps can deceptively extort vulnerable children by stylishly requesting for token fees for the registration of participants in order to facilitate logistics.

Figures 4–7 establishes the correlation between the elements of the crime and the circumstances that may surround vulnerable children to be alleged of juvenile crimes with malware apps. **Figure 4** basically reveals that the act of vindictive, disconsolate, failures, misfortune, castaway, down grading and downcast can create unwholesome and objectionable circumstances for vulnerable children to be alleged of juvenile crimes with mobile apps. **Figure 5** suggests that news of failure, parental death, planned impulses and unplanned impulses can further constitute derived determinants of the circumstances that can influence vulnerable children to be alleged of juvenile crimes with mobile apps.

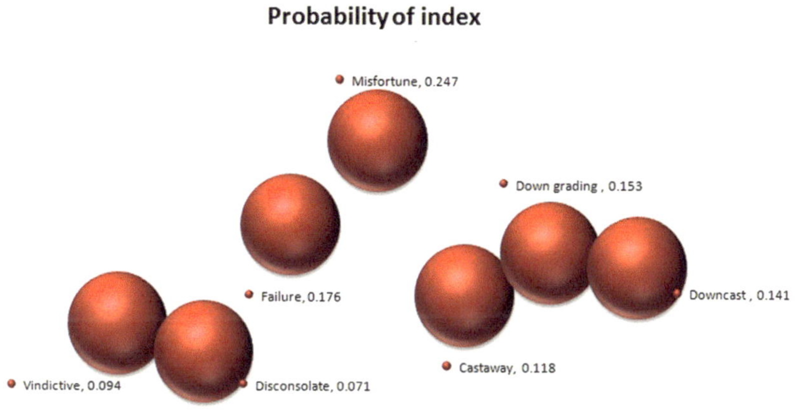

Figure 4.
First category of rationale that underlie the elements of juvenile crime with malware apps.

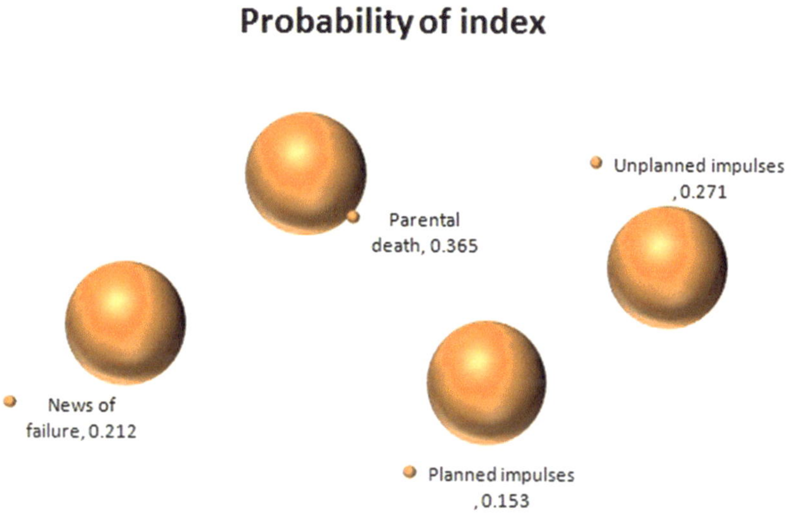

Figure 5.
Second category of rationale that underlie the elements of juvenile crime with malware apps.

The above observations indicate that indictive and disconsolate exhibit close statistical significance. Hence, they appear to cluster together. Conversely, failure and misfortune are not statistically close in significance. Hence, both factors appear to disperse from each other. Additionally, castaway, down grading and downcast have close statistical significance. Hence, these three factors appear to cluster together. The diagonal analysis of the variables suggest that parental death relatively aligns with news of failure as indicated by the closeness of their probabilities of occurence. Similarly, unplanned impulses diagonally aligns with planned impulses as supported by the closeness of their probabilities of occurence. These observations suggest that closely alligned variable can influence vulnerable children in almost the same way.

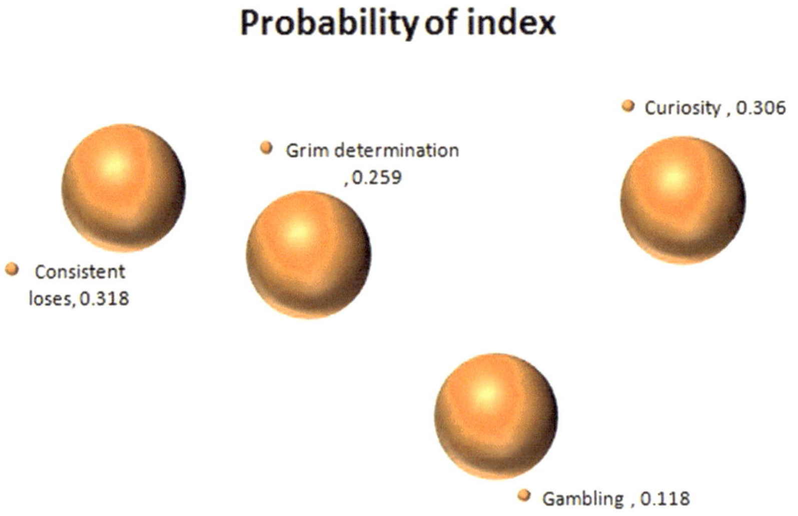

Figure 6.
Third category of rationale that underlie the elements of juvenile crime with malware apps.

Figure 6 believes that consistent loses; grim determination, curiosity and gambling are third elements of the crime that may create another group of unwholesome and objectionable circumstances for vulnerable children to be alleged of juvenile crimes with malicious apps. The results showed that consistent loses and grim determinations have close likelihood of occurrences. Hence, they appear to close to each order diagonally. In a contrast manner, gambling and curiosity are far apart because of the disparity in the likelihood of their occurrences. The diagonal analysis of these variables suggest that they do not relatively align with each other. These observations propose that grim determinations, consistemt failure, gambling and cusriosity can underlie the bahaviour of vulnerable children in different ways.

Figure 7 establishes that bankruptcy, discontentment, disappointment and antagonism are hidden factors that underlie the fundamental elements of juvenile crimes with malware apps. The closeness in the individual probability of occurrence of the above factors connotes that they are likely to regularly compound the above issues in creating another group of unwholesome and objectionable circumstances for most of the vulnerable children to be alleged of juvenile crimes with mobile apps. The above analysis shows that anger often correlates to discontentment in some children. The above empirical experiments suggest that disappointment that comes in the form of sudden frustration whereby a child is expecting hope but consistently turn out to be hopeless expectations can induce juvenile crimes committed with malware apps. Excessive domestic violence, hostility and actions of peers, families and schools that consistently antagonizing a child can significantly lure vulnerable children to commit juvenile crimes with malware apps.

Grim determinations describe a circumstance whereby it is impossible to plea, appease or to calm down a child. Disconsolate describes a circumstance whereby a child can be sad or dejected beyond comforting such that he/she could be incapable of being consoled by fellow human being. Vindictive is a circumstance whereby a child is determined or disposed to seek for revenge, avenge or intends to seek for revenge at all cost. Bankruptcy describes the circumstance whereby a child completely or suddenly get ruined or discovers that he/she has lacked some basic, moral, spiritual and

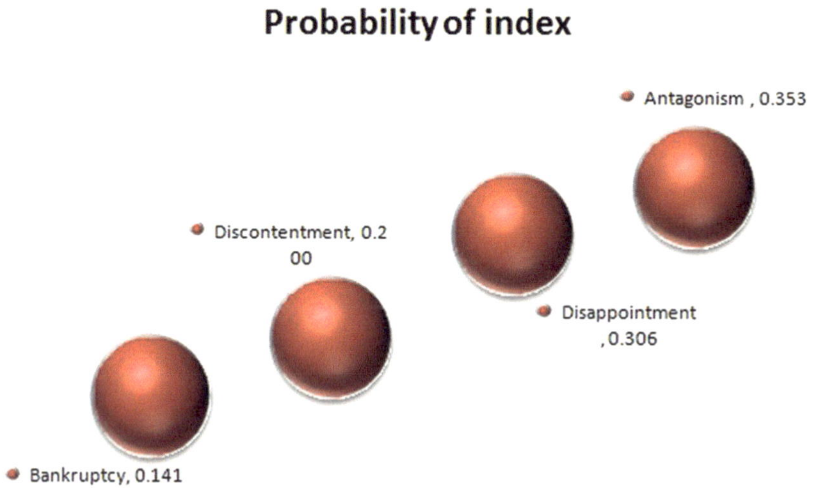

Figure 7.
Fourth category of rationale that underlie the elements of juvenile crime with malware apps.

intellectual properties that constitute wellbeing in human life [14]. Discontentment is the circumstance whereby a child is unhappy due to failure to meet up with his/her yearning desire, especially for the desire to acquire or attain something better than his or her present situation or attainment in the society.

Further still, according to the Oxford dictionary [14], disappointment is the circumstance whereby a child conceives the feeling of dissatisfaction especially if his/her expectations are not realized at the expected time. Dissatisfaction is a circumstance whereby a child is feeling of frustration or a child is being displeased (or perceives) disgruntlement with particular services, himself/herself or some things in the society. News of failure is a circumstance where a child is unable to manage sudden (or intermittent) news of consistent failure. This may occur if a vulnerable child is unable to successfully accomplish the projected height, tasks or goals in life having tried all his or her capability to accomplish the goals (tasks) on several occasions. Castaway is the circumstance whereby a child is rejected or discarded (or the child continues to feel that he/she is rejected or discarded) by his/her home and the society that he/she lives. Down grading is a circumstance whereby a child is reduced and blatantly made to appear (or continues to feel) as if he/she is an insignificant person in home and society. Down cast is a circumstance whereby a child is dejected or depressed (or the child feels that he/she is wretched) beyond immediate recovery. Misfortune is the circumstance whereby a child perceives bad luck, or the child is facing untold hardship or uncontrollable hard times. Antagonism is the circumstance whereby a child is often confronted with uncontrollable and inflexible opposition such as rigid hostility; lack of sympathy, resentment and firm hatred.

In reality, juveniles are young people that have not fully attained adulthood. They may depend on their parents, guardians or fostering cares to meet some basic requirements of daily living. They must not be abandoned so that they will not grow up with abandoned hope. Juveniles will need empathy, caring and loving people that can comfort and motivate them whenever they are contestants and whenever they suddenly stumble on obstacles (or unavoidable pitfalls) in life that can make them record sudden loses in competitions. Therefore, adequate social and restorative or rehabilitative interventions should be specially designed for children that are affected with sudden news of failure, parental death and depression [24].

The above findings have raised about five legal and technical discussions in modern society. Suppose a juvenile crime is committed with malware apps and by means of mobile phone that involves the illegal movement of huge cash from an account of a customer into an account in another bank. This illegal cash movement obviously involves two banks. Then, who should have stopped the crime? Should we attribute the "negligence to have prevented the crime" to the two banks, telecommunication or software companies that provide banking services to both banks? Who will the victim (bank's customer) sue in this case? Thus, the above striking observeations have also called for sudden review of social policies for young children to assist parenting and policing in adequately safeguarding vulnerable children across the globe. Governments should diversify social interventions towards the various forms of the derived determinants of juvenile crimes with malware apps that are discovered above. Media involvement in the education of masses on the dangers of malware apps, unintentional cybercrimes and how innocent children may be incriminated for unknowingly committing juvenile crimes with malware apps are urgently needed in all nations of the globe. Policy makers might need to review the enabling legislations to explicate new legal issues that we raised in this chapter. In effect, software companies and service providers that market or design computer apps must thoroughly review their Service Level Agreements (SLAs) together with the underlying security functionalities of their products to accomondate new legal concepts in software industry. For instance, the issues of transfer of criminal liability, shared or joint liability, criminal consent of underaged children and the other vital issues that we have identified above demands urgent review in law books and contemporay bulletins. Government should review and direct some social interventions towards enhancing mass literacy in order to discourage youths from kleptomaniac lifestyle and excessive passion to acquire money by all means. Social interventions that will enable children to strictly adhere to the standard way of live and regulation of the level of exposure of children to some social activities (e.g. films and video) must be vigorously emphasized.

Comprehensive suit that indicate package of social interventions such as legislative protection, scholarship and financial grants to indigent children are recommended for assisting vulnerable juveniles in the above context. Governments across the globe should constantly review the existing social interventions to ensure hat they are properly tailored towards the prevention of the above activities that may compel juveniles to be vindictive, disconsolate and experience unmanageable failures, misfortune, castaway, down grading and downcast in the society. Interventions should countering the negative impacts of news of failure, parental death, planned impulses and unplanned impulses and consistent loses. Juveniles require educative interventions to elinghten them on the dangers in grim determination, excessive curiosity and gambling and how they can manage unforseen circumstances in early life. The above thoughts and pradigms are strongly recommeded to countering the kinds of juvenile crimes that are studied in this chapter.

6. Conclusion and summary

This chapter has shown that malware apps are malicious apps that are strongly competing with legitimate computer apps in the software industry in recent time. It has been shown that malware apps have intrusive or criminally-minded motives that underlie their functions, usage and practices. The motives of their designers and their users are to use them to corrupt, explore, delete, deface, swindle or steal resources or personal details of another people. Thus, we further classified prominent malware apps that are competing with legitimate computer apps in the software industry on the basis of their coaching skills. Several social and legal dilemmas came to fore in this study. The rationale that underlies the behavior of a child may be influenced by many unexpressed and unnoticeable circumstances that may initially elude the imaginations of their parents, guardians and fosters. These determinants can gradually upsurge and eventually attain high climax over time. The danger begins to increase if the affected child voluntarily or involuntarily commits juvenile crimes with malware apps against known or known victims. The influx of software industry with numerous computer apps has led to the classifications of computer apps into two groups in this chapter. Many business-oriented apps are target of malware apps that have been designed by fraudulent people.

Moreover, bankruptcy, economic failure and poverty may be inducers of juvenile crimes committed with malware apps. This study further observes that some factors such as grim determination, disconsolate, parental death, downcast, disappointment; uncertainties about future and down grading a child, etc. that underlie the fundamental elements of the above crimes have not been empirically substantiated over the years. In other words, the correlation between the elements of the crime and the circumstances that surround vulnerable children that were rarely buttressed and made explicit by empirical studies over the year have been empirically investigated in this chapter. For this reason, virtual interactions with the participants in the survey that is reported in this chapter adduce the above lapses and their correlations with the dwindling of the efficacies of several social policies in most countries. Some of the intentions of malware apps like deceptive coaching, deceptive dating guidance and fake financial coaching tips have been enumerated above. We specifically argue that malware apps can expose the users of legitimate apps and continued usage of legitimate apps to greater risk of distraction and sudden neglect if the trend of their incursion into the software industry is not quickly curtailed.

Above all, this chapter has raised novel legal debates on four vital paradigms and other technological issues concerning the legal and technical interpretations of juvenile crimes committed with malware app that urgently call for in-depth legal consideration and interpretations. We submit that malware apps with deceptive motives may send malicious files (or documents) to a vulnerable child and lure the child to download (or open) the files in order to unlock his/her internet account. The files may be Trojans or virus that will in turn corrupt other computer apps in the digital systems of the child or flout the security of another person without the consent of the child. We have therefore identified new issues regarding shared or joint criminal liability, the transfer of criminal liability, the restrictive applicability of the criminal consent of a child that is still under the tutelage and guidance of their parents and the circumstances whereby some legitimate apps may be held (or charged) by complainants for being liable to have criminally permitted some juvenile crimes that severely impact them (or to have permeated malware apps that spread virus or malicious information to propagate through them either by accidental (inadvertent, unintentional or unplanned) or intentional manner.

In addition, another dilemma that we put forward in this chapter is how the software companies and service providers can acceptably account for fake voice apps (for instance) that are proven to have adopted the data (e.g. multimedia data and textual data) they have previously gathered from vulnerable children to defraud some people or swindle some vulnerable friends or cause controversies among the relatives of the users in the later time. We therefore raise new legal debates regarding industrial conflict and amicable resolution of disputes in relation to the procedures for compensating end-users (or paying) for the damages end-users might have incurred by virtue of "the services they receive and the confidence (or trust) they have that underlie the use of the computer apps", especially if the damages are proven to be directly caused by the occurrence of unexpected intrusions that the manufacturers of the computer apps or service providers (vendors) should have stopped from taking place.

The premise is that the impacts of juvenile crimes with malware apps on victims must not statistically correlate to stigmatization, shame, stress, hurtful experience or increased anxiety. Otherwise, several victims of the crimes may cover up their tribulations and harms that they have incurred from the crimes. Therefore, we suggest that global governance should be directed towards the review and constant improvement of criminal laws to ensure that local and multinationals together with third party (service providers) can be effectively held to account for aiding or failing to mitigate serious juvenile crimes committed with malware apps and their variants. We also suggest that global citizens and media sector should work together to advocate and stand against all forms of discriminations against victims of the above crimes. We solidly believe that victims of juvenile crimes with malware apps require social interventions for them to recover to normal life. We recommend that future research work should delve into the above legal paradigms. Inter-disciplinary collaborations that will ensure accurate media's representations of the methodology, prevalent and prevention of juvenile crimes that some children may commit with malware apps are inevitable in order to enlighten vulnerable children and to maximize the efficacies of the preventive and restorative interventions that governments and Non-Governmental Organizations (NGOs) have designed for the above category of global crimes.

Finally, we believe that the lists of legitimate apps and malware apps in the software industry across the globe may be inexhaustible when compare to all the computer apps that we have mentioned in this chapter. There are possibilities of fake homework coaching apps and fake fitness coaching apps. We might have only discussed malicious apps that may not comprise the entire competitors with legitimate apps in the industry. Therefore, we strongly recommend future research work to improve on the limitations of the above research findings.

Author details

Joshua Ojo Nehinbe
ICT Security Solutions, W/Africa

*Address all correspondence to: nehinbe@yahoo.com

References

[1] Clarke-Midura J, Sun C, Pantic K. Making apps: An approach to recruiting youth to computer science. ACM Transactions on Computing Education. 2020;**20**(4):1-23. DOI: 10.1145/3425710

[2] Wright JH, Mishkind M, Eells TD, Chan SR. Computer-assisted cognitive-behavior therapy and mobile apps for depression and anxiety. Current Psychiatry Reports. 2019;**21**:62. Available from: https://pubmed.ncbi.nlm.nih.gov/31250242/

[3] Aditya K. Comparative Study of Juvenile Delinquency Law Between India, USA and UK. 2022. Available from: https://ssrn.com/abstract=3607875 [Accessed: May 22, 2020]

[4] Gatti U, Tremblay R, Vitaro F, McDuff P. Youth gangs, delinquency and drug use: A test of the selection, facilitation, and enhancement hypotheses. Child Psychology and Psychiatry. 2005;**46**(11):1178-1190. DOI: 10.1111/j.1469-7610.2005.00423.x

[5] Davis AM. Great Software Debates. Wiley-IEEE Computer Society; 2004. ISBN-13: 978-0471675235

[6] University of Minnesota. Criminal Law: Element of Crime. USA. Available from: https://open.lib.umn.edu/criminallaw/chapter/4-1-criminal-elements/: University of Minnesota; 2022 [Accessed: July 31, 2022]

[7] Omoniyi MBI. Juvenile crimes and its Counseling implications. Journal of Psychology. 2011;**2**(1):1-6. DOI: 10.1080/09764224.2011.11885455

[8] National Research Council and Institute of Medicine (NRCIM). Juvenile crime, juvenile justice. Panel on juvenile crime: Prevention, treatment, and control. In: McCord J, Widom CS, Crowell NA, editors. Committee on Law and Justice and Board on Children, Youth, and Families. Washington, DC: National Academy Press; 2001. DOI: 10.17226/9747

[9] Eastcom C. Mathematically modelling victim selection in cyber crimes. In: ICCWS 2021 16th International Conference on Cyber Warfare and Security. USA: Tennessee Tech University and Oak Ridge; 2021

[10] Wu J, Hu X, Orrick EA. The relationship between motivations for joining gangs and violent offending: A preliminary test on self-determination theory. Victims & Offenders. 2022;**17**(3):335-349. DOI: 10.1080/15564886.2021.1898508

[11] Lee Y, Tae SG. A modeling perspective of juvenile crimes. International Journal of Numerical Analysis and Modeling. 2011;**2**(4):369-378

[12] Oxford Dictionaries. Crime. 2022. Available from: https://www.oxfordlearnersdictionaries.com/definition/english/crime?q=crime. [Accessed: August 12, 2022]

[13] Oxford Dictionaries. Commission. 2022. Available from: https://www.oxfordlearnersdictionaries.com/definition/english/commission_1?q=commission. [Accessed: August 12, 2022]

[14] Oxford Dictionary. Disappointment. 2022. Available from: https://www.oxfordlearnersdictionaries.com/definition/american_english/disappointment. [Accessed: September 14, 2022]

[15] Shamim A, Batool Z, Zafar MI, Hashmi N. A study of juvenile crimes

in Borstal jail, Faisalabad, Pakistan. The Journal of Animal & Plant Sciences. 2009;**19**(2):101-103

[16] Young U, Greer B, Church R. Juvenile Delinquency, Welfare, Justice and Therapeutic Interventions: A Global Perspective. London: Cambridge University Press; 2018

[17] Ellis L, Beaver K, Wright J. Handbook of Crime Correlates. 2nd ed. Academic Press; 2019. ISBN: 9780128044773

[18] Razzaq A, Hur A, Ahmad HF, Masood M. Cyber security: Threats, reasons, challenges, methodologies and state of the art solutions for industrial applications. In: 2013 IEEE Eleventh International Symposium on Autonomous Decentralized Systems (ISADS). 2013, 2013. pp. 1-6. DOI: 10.1109/ISADS.2013.6513420

[19] Rajib M. Fundamentals of Software Engineering. 5th ed. Delhi, India: PHI Learning Pvt. Ltd.; 2018

[20] Qi-qi H. Analysis on the causes of juvenile crimes from the perspective of psychology. Academic Journal of Humanities & Social Sciences. 2020;**3**(10):55-59. DOI: 10.25236/AJHSS.2020.031008

[21] Saylordotorg. Chapter 4 The Elements of a Crime. 2022. Available from: https://saylordotorg.github.io/text_criminal-law/s08-the-elements-of-a-crime.html. [Accessed: July 30, 2022]

[22] Britannica. The Elements of Crime. 2022. Available from: https://www.britannica.com/ topic/criminal-law/The-elements-of-crime. [Accessed: July 20, 2022]

[23] Brittany McKenna Show bio. The Elements of a Crime: Definition & Overview. 2021. Available from: https://study.com/academy/lesson/the-elements-of-a-crime-definition-lesson.html. [Accessed: July 30, 2022]

[24] Jufria M, Nazerib NBM, Dhanapal S. Restorative justice: An alternative process for solving juvenile crimes in Indonesia. Brawijaya Law Journal. 2019;**6**(2):157-169

Section 2

Malware Detection

Chapter 3

Detection and Minimization of Malware by Implementing AI in SMEs

Nisha Rawindaran, Liqaa Nawaf, Vibhushinie Bentotahewa, Edmond Prakash, Ambikesh Jayal, Chaminda Hewage and Daniyal Mohammed N. Alghazzawi

Abstract

The malware can threaten personal privacy by opening backdoors for attackers to access user passwords, IP addresses, banking information, and other personal data, whilst some malware extracts personal data and sends them to people unknown to the users. In this chapter, the authors will present recent case studies and discuss the privacy and security threats associated with different types of malwares. The small medium enterprises (SMEs) have a unique working model forming the backbone of the UK economy and malware affects SMEs' organizations. Also, the use of Artificial Intelligence (AI) as both an offense and defense mechanism, for the hacker, and the end user will be investigated further. In conclusion, finding a balance between IT expertise and the costs of products that are able to help SMEs protect and secure their data will benefit the SMEs by using a more intelligent controlled environment with applied machine learning techniques and not compromising on costs will be discussed.

Keywords: malware, privacy, cyber defense, ransomware, SME, artificial intelligence, machine learning, big data, GDPR

1. Introduction

The people in large numbers continue to keep pace with rapidly evolving technology and that has led to increased use of computers and mobile devices. This trend was clearly visible during the COVID-19 pandemic when traditional businesses resorted to using modern technologies and became dependent on applications such as QR codes and contactless payments. Against that background, the hackers took advantage of the prevailing circumstances to exploit the vulnerabilities of technology and systems, and the level of risks to privacy and national security became a serious concern to the respective authorities.

This environment became hackers' paradise and directly or indirectly provided them with an ideal opportunity to steal personal data and sell them to third parties in return for financial benefits. They targeted large companies as well as Small and Medium-sized enterprises (SMEs). The SMEs switched to new technologies to maintain their business during the COVID-19 pandemic, but they invariably failed to increase the security aspects of the new systems due to financial constraints. Some researchers have pointed out that well-prepared organizations were able to deal with cyber incidents more efficiently than those that failed to anticipate and plan to address the reality of cyber threats due to the lack of adequate capabilities [1]. Statistically, SMEs represent more than 99% of all businesses in the UK, and given the importance of this commercial sector, the impact of disproportionate financial resources on the operational and reputational of the SMEs is of serious concern [2]. It is therefore crucially important to understand the threats early and act on them to prevent unimaginable repercussions.

SMEs have a unique working model forming the backbone of the UK economy. According to the Federation of Small Businesses (FSB) UK business statistics, there were 5.5 million small businesses at the start of 2021 [3]. The statistics showed SMEs accounted for 99.9% of this business population making up three-fifths of the employment and half of turnover in the UK private sector. These statistics verify how important the SME ecosystem is in providing an important cog to the growth and economy of a developed country. SMEs have an important role to play begging the question of how their usage of emerging technologies is keeping their data and business safe online. A paper by Daniel and Andreas [3], explores these emerging technologies especially the use of Artificial Intelligence (AI) and Machine Learning (ML) as both offense and defense mechanisms, for the hacker, and the end user. Daniel and Andreas identify and evaluate AI-related use cases that have a high impact potential on the cyber security level of SMEs, in particular highlighting the challenges of SME's environment being low in resourcing and challenges in their financial capabilities. AI and ML can be utilized for the defense against cyber threats especially malicious software (Malware). Attacks can easily be obtained from the dark web via malware-as-a-service (MAAS) making the underworld choices easier to conduct. Hackers with limited knowledge are able to use AI technology in order to create chaos and havoc in cyberspace. Traditional signature-based security systems can detect only 75–95% of untargeted mass malware attacks compared to 27% of targeted malware cyber-attacks [4] according to Daniel and Andreas. The detection rates of IT systems that do not use any form of AI cannot be sustained at the same level of security and protection when attackers are also using modern levels of AI methods to attack IT systems.

According to a study by Rawindaran et al. [5], over one million new types of malwares are created each day by malicious hackers. These types of malwares try to infiltrate networks increasing the threat of network attacks, driving the usage and demand for the use of AI and ML-driven intrusion detection protection systems (IDPS) being used throughout the SME market. These systems come with many challenges that include the cost of buying and maintaining the system and resourcing skilled engineers to maintain these systems in order to create a healthy and safe environment within their business [5]. Rawindaran et al. also explored a cost model to understand the outcome of SMEs' decision-making, in getting the right framework in place in securing their data. An experiment was conducted comparing different software vendors in understanding the information captured using AI and ML technology to stop zero-day attacks. The requirements of the UK General Data Protection Regulations Act (GDPR) were also acknowledged as part of the broader framework of the study. ML techniques such as anomaly-based intrusions did show better detection through a commercially subscription-based model for support from Cisco compared to that of the Open Source model which required internal expertise in ML. Finding a balance between IT expertise and the costs of products that are able to help SMEs protect and secure their data, will benefit the SMEs from using a more intelligent controlled environment with applied ML techniques, whilst not compromising on costs. This research work also focuses on evaluating techniques for managing big data and detecting malware within SMEs.

2. Related work

There is a wealth of literature covering many aspects of malware impact, including data protection and privacy, big data risks, and AI defense mechanisms as discussed in the following subsections. In this chapter, we restrict our attention to the areas relevant to malware detection in SMEs by implementing AI.

2.1 Data protection and privacy

Data protection against malware is the biggest challenge facing the industry, and in general, malware represents a classic example of a confidentiality breach exposing personal data held by a company on its servers [6]. Trojan Horse Virus is a type of malware when downloaded to a computer resembles a legitimate program, the hacktivists use social engineering to infect computers with it [7]. Most Trojans are designed to take control of a user's computer to steal data and feed more malware in the process, and when once a host computer is infected, they pose significant threats to the business. However, they do not particularly target large organizations, and according to the reports, 29% of cybercrimes affecting SMEs were caused by malware attacks [8]. The Trojan viruses rely on human error [9], and unlike other forms of malware, they do not simply appear on a machine and start damaging the system on their own, and need to be manually opened, when triggered [9] have the capability to delete, block, modify, copy data, and disrupt the performance of the computers [10] by causing damage to the organization's data and personal data collected, processed, and stored in them.

Phishing is another variant of cyber-attack used by the hacktivists to victimize users by unethically persuading the victims to disclose personal and other critical information, and they do this by asking the user to perform standard procedures in personal data handling such as clicking on a connection to download files and applications [11]. The attackers use this technique to transmit malicious links containing viruses as worms, and if the victim followed the given instructions, the attacker would have access to private systems and personal information held in them. The staff can easily be caught unaware of the dangers due to a lack of knowledge and slackness in concentration and leaking personal data to the outside world would cause immense damage to the organization with serious repercussions. In the period between 2013 and 2015, in an extended phishing campaign, the attacker sent faked invoices impersonating Quanta (a Taiwan-based company) as a vendor, to Facebook and Google and both companies were tricked out a sum of $100 million in payments [12]. Eventually, Facebook and Google discovered the fraud and took legal action through the US legislature [12]. This clearly indicates that large companies are not immune, and the studies reveal that phishing attacks are among the most common cyber incidents that SMEs are likely to be affected and fall victim to [1]. For instance, the cybercriminals made concerted efforts to compromise accounts by using phishing emails with the subject 'Covid-19' [1], taking advantage of the concerns arising from the pandemic.

According to the National Cyber Security Centre (NCSC), ransomware is one of the most immediate dangers to UK businesses and other organizations [2]. 'Ransomware' is a particular variant of malware that collects data and network devices, and encrypts them, preventing user access [13], and access can be restored only by agreeing to the hacker's demands for paying the ransom. However, it is not always possible to regain access to locked data as the hacktivists can refuse to unlock the devices until the ransom is paid, or even after making the payment; in such a scenario, the organizations will incur data and final losses, whilst also end up with a damaged reputation in the eyes of the public. Hackers have targeted a range of industries, Automotive, Business services, Food and agriculture, Healthcare, Insurance, Law enforcement, Oil and gas, and Tech, demanding ransoms [14], and SMEs have not escaped harassment from the criminals. The reports suggest that many SMEs tend to assume that 'higher value targets' such as critical infrastructure and larger organizations, are likely to be prime targets [2], but the statistics suggest that 82% of Ransomware Attacks also target Small Businesses [15]. The inference drawn from these statistics is that the primary objective of hackers is to obtain data regardless of the size of the organization. The reports also suggest that, in the recent climate of cyber incidents, ransomware attacks have become a menace to organizations as well as to individuals hindering timely access to their personal data [16]. Therefore, any organization collecting, processing, and storing information should stay alert to the threats.

Another malicious software is known as 'Spyware, which is a malicious software/ apps (malware) stealthily installed to monitor and track device activity [17]. It allows the attackers remote access to the victim's devices, and it enables the hackers to invade the privacy of the people by reading messages, listening to phone calls, accessing photos, viewing browsing history, capturing, and transferring audio and camera recordings in real-time [18]. According to the statistics, spyware is the third most popular malware used in attacks against organizations in 2021, and it is the second most used in attacks on individuals [19]. In January 2020, a United Nations (UN) investigation discovered that the Amazon CEO's (Jeff Bezos') smartphone was targeted by spyware and several megabytes of data was extracted from it over a considerable period (months) [19]. The UN report identified Pegasus spyware, which was created and sold by Israel-based NSO Group, as the intruder, and another investigation conducted by Amnesty International's cybersecurity team identified the same spyware as the intruder found in the phones of hundreds of people [19]. Therefore, those organizations dealing with personal data need to be aware of the reality of the threats and have preventative mechanisms to deter spyware attacks. The failure to do so will allow hackers to obtain sensitive data and sell captured data through spyware attacks on the Dark Web, and consequences leading to immense damage to personal privacy in the first place, and eventually with threats to national security.

Adware is another (unwelcome) software designed to throw advertisements up on your screen, and they piggyback on another program to trick you into installing it on your PC, tablet, or mobile device [20]. Once the user's device is hijacked by the adware, it detects the location, collects information about the Internet sites visited, and presents advertising pertaining to the types of goods or services searched by the user [20]. There is also a risk of data being shared with third parties, and that amounts to a violation of personal privacy. The threat posed by adware is not limited to large companies, regardless of the size of the organization it affects everyone, and the consequences can range from threats to personal security as well as to national security.

Malware, in general, represents a classic example of a confidentiality breach extracting and exposing personal data held by a company by hacking and downloading personal data from systems and devices [6]. In the process, malware cause interruptions to the network of the enterprise irrespective of the organization's size. Malware also has the capability to record browsing history, monitor applications being used, and make copies of personal information like user IDs, passwords, and bank account details. That is not all, Malware by hacking the network of an organization can affect the confidentiality and integrity of personal data and delete/edit/steal personal data. In some cases, malware can potentially disable critical services offered by the company, and that will make the services unavailable to the clients with consequences damaging the image of the organization's reputation, affecting trustworthiness, and contributing to financial losses.

Data is valuable to any organization whether it happened to be large or small, and the crucial issue is that it is a sellable item and anyone getting access to it can make money by selling it to the highest bidder on the dark web. Data is wide-ranging and consist of information about the organization, and sensitive personal data about the employees and the clients, and the onus is on the respective organization to ensure the security of that data by having in place adequate data protection mechanism in compliance with data protection regulations applicable at the regional or country level. For example, an organization in the UK collecting, processing, and storing information about their customers, has an obligation to follow data protection mechanisms/guidelines set out in UK GDPR that is on par with the EU regulations. Therefore, it makes any organization outside the EU engaged in commercial activities processing information of the citizens of the EU also bound by EU GDPR.

The application of the GDPR does not depend on the size of the organization. Whether it is an SME or a large organization in the EU or UK, if they collect, process, and store data, they should abide by the GDPR regulations. However, some of the obligations of the GDPR may not apply to all SMEs with less than 250 employees. For example, SMEs do not have to keep records of their processing activities unless the processing of personal data is a regular occurrence, poses a potential threat to individuals' rights and freedom, or includes sensitive data or criminal records [21, 22]. Also, SMEs are required to appoint a Data Protection Officer only if the organization is processing data as part of the main business, and it may also pose threats to an individual's rights and freedoms [22].

A common misconception was that the GDPR would only be looking into the data protection practices of large multinational enterprises. The €50 m fine imposed on Google by Commission Nationale de l'informatique et des libertés (CNIL) or the €204 m fine imposed on British Airways were high in comparison to what had been imposed on smaller enterprises [23]. However, a CNIL had imposed a fine of €400,000 on the real estate firm, SERGIC, whilst the less performing advertising Agency, QuickClickNow, was served with a fine of only €47,000 by the Polish Data Protection Authority [23].

GDPR stipulates that data breach of any kind associated with any variant of malware attack, the data subject, and the relevant authorities should be notified within 72 hours [6], and data breach notification should include the details of the nature of the breach. These are specified as personal data, the name and contact details of the data protection officer, contact point for obtaining additional information, consequences of the personal data breach; description of the measures taken or proposed to deal with the data breach, and description of the measures taken to mitigate any adverse effects, clear advice on the steps that the individuals should take to protect themselves, and what assistance the organization would be prepared offer them [24] This framework of data protection mechanisms would provide the organizations the knowhow and competences to deal with malware-related data breaches with confidence and to avoid the damage and impact on the reputation and trustworthiness of the organization.

2.2 Malware impacts on SMEs

Every year a survey of the UK Cyber Security Breach Survey 2022 [25] is conducted by the Department of Digital, Culture, Media, and Sports (DCMS) in line with the UK National Cyber Strategy, detailing the cost and impact of cyber-attacks on UK businesses. The year 2022 showed that 39% of UK businesses have been a victim of cyber-attacks. The most common attacks were phishing which accounted for 83 and 21% were attack types such as a denial of service, malware, or ransomware attack. Despite having a lower percentage of attacks, businesses cited ransomware as a major threat, with 56% having a policy not to pay the ransom as quoted by this survey. The report gave insight on SMEs and the culture of not believing ransomware will pose a threat to their business, and that it was unlikely to happen, as SMEs assumed that they did not hold anything of value the hackers would be interested in. In the event of a cyber-attack, the SMEs in this survey tended to resort to traditional methods of recovery, such as shutting down systems and re-booting with backed-up data. SMEs also sought advice from external IT providers for assistance in the event of an attack and some had no plans at all. However, SMEs who took cyber-attacks seriously viewed ransomware such as malware as a serious threat and often had a strict plan in place in the event of an attack. The survey particularly raised points such as SMEs not necessarily engaging in industry standards to help protect their business such as Cyber Essentials and how the uptake for this standard was still exceptionally low as they felt they did not meet the criteria. SMEs who felt they did not have the technical understanding was clearly observed in this survey to suggest that the role of external IT companies is particularly important in the supply chain model. This is so SMEs can access the benefits of a larger and more resourced specialist. The survey also highlighted the importance of how the supply chain poses a threat as an entry point for attackers and the business can only be as strong as the weakest link of the supply chain. The survey indicated that fewer than one in ten organizations actively monitor the risks within their supply chain and so this presents a clear risk for the future. This DCMS survey is particularly important as it is a window into the activities year on year on how trends of cyber-attacks affect businesses in the UK and how our behavioral patterns have a consequence towards risk and its management of it.

A study conducted by Tirumala et al. [26], explored that ransomware (malware) attacks have forced businesses to think about the security of their resources due to SMEs not having the right cyber defense mechanisms in place. This research explores implementing a "Raspberry Pi" based intelligent cyber defense system (iCDS) for SME networks and Smart-homes and how these devices are used to filter malicious contents from incoming traffic and be able to detect malware using AI. The paper concluded that the "Rasberry Pi" device is feasible to use to develop a low-cost iCDS as an alternative to the traditional rule based IDSs in use. Tirumala's study reinforces the study of Rawindaran et al. [5] on the uses of open-source IDS versus commercial IDS, and the challenges faced when introducing ML and AI technologies versus traditional

IDS to promote better hygiene within the cyber interface. Rawindaran et al. took requirements from the UK General Data Protection Regulations Act (GDPR) as part of the broader framework of the study and further explored the techniques of ML to show better detection through a commercially subscription-based model for support from Cisco compared to that of the Open-Source model which required internal expertise in ML. The study went on to discuss the challenges between IT expertise and costs of products to help SMEs protect and secure their data and the benefits of moving to an intelligently controlled environment and not compromising on costs. Kshetri [27] added that.

> *"Cybersecurity company Blue Voyant's survey found that 97% of firms had been impacted by a cybersecurity breach in their supply chain, and 93% had suffered a direct cybersecurity breach due to their supply chains' weaknesses." [27].*

There are various points in an SME business whereby malware can make its presence, and Kshetri's study performs an exploration into the various elements that contribute to the health of SMEs. Part of the vulnerability lies in the much-used supply chain partners to SMEs. These are in the form of third-party software, managed service providers (MSPs), IT vendors and other providers of software and its content, vendors in a physical capacity, and non-IT contracting vendors. For third-party software, the vulnerability lies in the "implanting" of malicious codes within this software. Understanding where this software come from and how they are managed within the supply chain can be a challenge and barrier. For MSPs, it is the reliance of these MSPs pushing out updates that could contain malicious code from their supply chain, in providing their own service of performing remote monitoring on managed computers. When it comes to IT vendors and partners, both virtual and physical, understanding the vulnerabilities of installing or injection of malicious codes by the attack on these vendors before products get shipped or provided to businesses. Lastly non-IT contracting vendors are using this platform to gain access to privileged resources to target the business. On each story of this supply chain, Kshetri gave examples, such as the attack on Equifax in 2017, showing a compromise of 146.6 million social security numbers and personal data. In the attack on Kaseya in 2021, victims were from 17 countries targeting nearly 1500 businesses. SolarWinds breached in 2020, showing an impact to 18,000 of their customers in the installation and injection of malicious codes. Most recently in the August of 2022, Advanced, a company that provides software for the NHS, experienced a ransomware attack causing patient data to be the target. Advanced as a supply chain provided NHS with services that included patient referrals, ambulance dispatch, out-of-hours appointment bookings, mental health services, and emergency prescriptions as reported by The Guardian [28].

Research is consistently showing ways in which to overcome malware attacks from various angles of business management, technology management, and of course human awareness within the SME. A paper by Cruzado et al. [29], suggested SMEs develop a "HOGO" reference framework based on two regulations, ISO 27002, and ISO 27032 for cybersecurity. The "HOGO" framework in this study applies good practices relating to internet security, critical infrastructures for information, network security, and information security, covering all aspects of the SME business [29]. Cruzado explains the framework being a combination of ISO 27002, which provides good controls for information security, and ISO 27032, providing good practices and recommendations. Both take into account the risks of the context for the security of the company's information. Using regulations can help identify supply chain partners with high cybersecurity risks (e.g., the U.K.'s Cyber Essentials) and reduce vulnerability in third-party software through frameworks such as "HOGO."

2.3 Big data risks

This section of the research is related to the essentials of the big data collection process through several resources from Wireless Sensors Networks (WSN) and other IoT sensors. Data collection is a procedure of collecting significant information to evaluate the outcome process and it becomes gradually significant since the burst of big data and the new development of technologies. SMEs collect an increasingly large amount of data, with information following into departments from many directions. For the data that SMEs collect to be meaningful and actionable, it needs to be provided in real-time so policy makers or managers can make decisions based on understanding the situation as it is, and not as it was. Thus, what technology is needed to make the most up-to-the-moment besides developing and modifying the policies to make the most of up-to-date data?

Big data refer to collecting and managing data in three forms High Volume, High Velocity, and Wide Variety. Big data management refers to the effective procedure that focuses on the management and usage of structured and unstructured data and its main purpose is to attain great data quality and accessibility for big data applications that certainly influence the performance of the organization [30]. The appropriate oversight of data throughout its life cycle is important to optimize its utility and minimize potential errors.

The most crucial purpose of big data is to guarantee that the data is captured and stored securely from resources, so it includes good data protection to avoid cyber risks. Big data management is challenging, and it has a vital role in managing the organization's data, it's a useful technique companies follow to maintain or preserve the data. The critical role in exploring and analyzing a big quantity of data is to discover effective patterns for big data. The business organization/SME aims to generate products and provide insight from this big data to improve its product achievements.

Recent technological developments in the field of communication struggle with internet connectivity issues that have led to the development of Wireless Mesh Networks (WMN). WMN is a wireless form of communication that works on the multi-hop concept for connecting multiple devices in the same grid area [31].

The multi-hop nature of WMN tied with fewer security mechanisms being employed makes it mandatory to make WMN secure from foreign attacks and malware. It's clear that security needs more attention in the characteristic of WMN. The Wireless Sensors Networks (WSN) enable the communication between devices and Radio Frequency Identification (RFID) allows the category of devices to collect the data. The amount of data collected from WSN puts entities at risk as they become more easily recognized with unauthorized processing, which can disrupt data protection laws, for instance, General Data Protection Regulation (GDPR), any data breaches, the data controlled will pay fines under GDPR (4% of annual turnover or 20 million) or evolving data protection laws [32]. There are many security concerns related to wireless communication, network transmission, information processing, and privacy. Also, two types of security parameters must be upgraded such as encryption and authentication.

The perception of business organizations/SMEs believes that more data is collected to gain visions and offer greater knowledge and greater benefit to the organizations, and data minimizations will limit the success of some specific applications.

Also, big data has a big impact on security performance and should be evaluated. Apparently, gathering mass amounts of data using WSN can be acceptable only if the benefits overweight the privacy and security of personal data. Therefore, securing personal data has become a significant challenge in contradiction to the growing malware and data risks [33]. There is more to be done in this aspect of the collected data addressing privacy and security concerns as explained in the previous Section 2.1.

Technology development is predicted based on the collected data from a particular application, so examining the collected data and detecting any deviations to report the error is significant by applying artificial intelligence, such as the machine learning algorithm that will help to perform and detect malware. Prediction is made by different data mining approaches using the data set through the networks. Sophisticated algorithms are mainly used to predict and detect malware. Further investigation of the potential malware risk is recommended to optimize security in SMEs.

2.4 Artificial intelligence for defense mechanism

There are high risks with big data in SMEs and it is crucial and significant to preserve by determining the security and utilizing a secure protocol that could be the contributing factor prevent various types of attacks. A novel method by applying a metaheuristic algorithm would be suggested for security and protection. Metaheuristic algorithms are general-purpose algorithms that can be applied to a wide range of optimization problems, with only minor alterations and modifications to the basic algorithm definition. Most metaheuristic techniques attempt to mimic biological, physical, and natural phenomena. Many heuristic and metaheuristic algorithms have been applied to improve solutions quality and solve large complex network optimization problems of maintaining QoS and have been shown to be important tools in a variety of disciplines. Metaheuristic methods can be developed to determine the best location to place the infrastructure and data to optimize security and reduce risks arising [34, 35].

The conversations and collaboration have taken place around making sure the necessary infrastructure is put in a secure place to help these SMEs succeed. The world is becoming smaller and smaller, so we need to bet on these digital innovations and help SMEs to reach out to secure markets where there is no longer a traditional definition. It is also important to consider the environmentally sustainable financial issues that are built to support these sectors. The overall vision is to optimize security and reduce malware risk in SMEs based on the recent technological revolution using AI approaches.

3. Technical methodology

The research methodology describes the methods used to collect and analyze data. The selection of the data collection method depends on the nature, scope, aims, and objectives of the research. In this research, the authors collected the data using secondary and primary data collection methods and analyzed the data to answer the research question. The outcome of this research aims to add new knowledge to the existing literature. The authors used published data in books, government publications, newspapers, magazines, and journals. These sources provided factual data related to the research scope. Finding secondary data with 100 percent applicability to one particular research scope is difficult; therefore, the primary data collection method has also been used in this research. However, the data collected from secondary resources have credibility and contribute to the validity of the research study.

The technical methodology took the form of a survey questionnaire. In order for the authors to deeper understand how SMEs, manage the detection and minimization of malware, the implementation of AI over standard traditional methods are discussed in this paper. The aim of this paper is to understand SMEs' awareness of Machine Learning Cyber Security (MLCS) software packages in SMEs; hence a survey was conducted to observe the uses of MLCS in the protection of data within the SME, against cyber threats as part of the SME cyber security implementations. A survey questionnaire was conducted through market research to a targeted SME audience in Wales. The survey was sent to 600 registered members of the Cyber Resilient Centre (CRC) of Wales. The survey was sent via email. The results showed that 40 people completed the survey with completed answers. The majority of the questions were answered by SME management between the ages of 46–55 and being decision-makers within the business. As part of GDPR, participation was voluntary, and all responses were kept anonymous. Consent was obtained and participants were given the right to withdraw and cancel participation at any one time. Various questions were asked to the SME population members of CRC.

4. Discussion and implemented techniques

The results from the survey are discussed further and analyzed to give feedback accordingly. **Figure 1** below showed that 85% of the respondents came back stating that they have the right cyber security software package in place to protect their business from cyber threats. The percentage who did not have any software in place was 15% of the respondents as shown below in the pie chart.

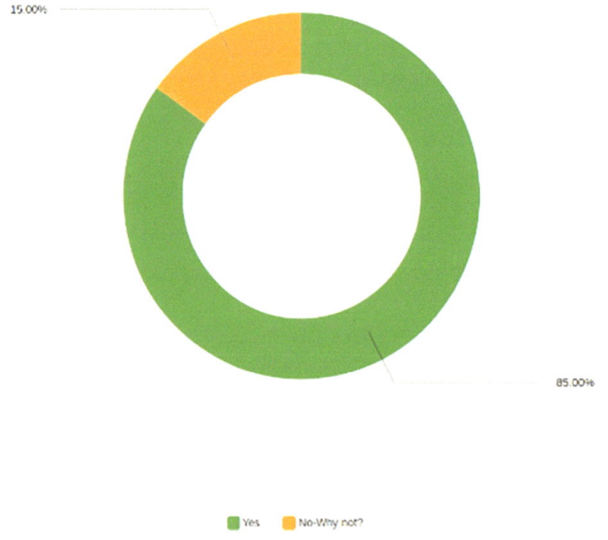

Figure 1.
SMEs having the right cyber security software package.

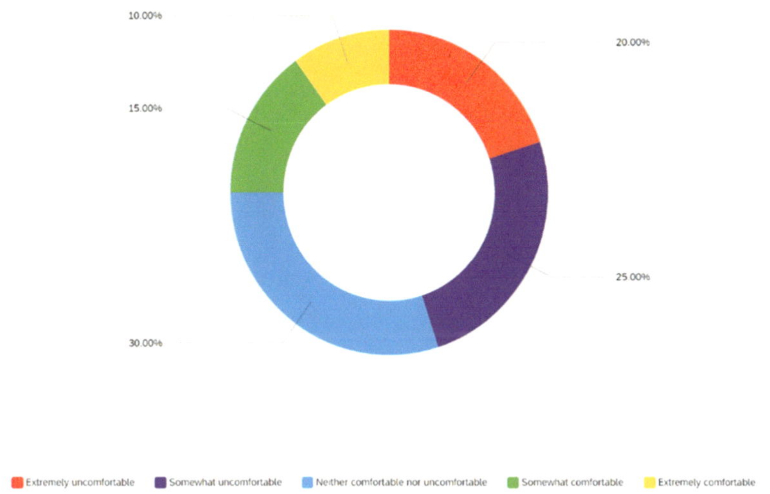

Figure 2.
SMEs understanding of ML.

From the 85% of respondents that said "Yes" in **Figure 1**, used a combination of commercial, non-commercial, and open-source software to protect their business. The next question looked at how comfortable SMEs were with the understanding of ML. **Figure 2** shows this breakdown with 10% being "Extremely comfortable" with ML, 15% being "Somewhat comfortable" and 30% being "Neither comfortable nor uncomfortable" with the understanding of ML.

Figure 2 went on to explain how 25% of SMEs were "Somewhat uncomfortable" with ML and 20% were "Extremely uncomfortable" with the understanding of ML. **Figure 3** answered the question on the awareness of MLCS and its application within the SME business. Nearly 11% were "Extremely Aware" of MLCS software packages within the business, 15% were "Yes, very nearly Aware" and 26% were "Somewhat Aware".

Figure 3 resulted in 47% of the SMEs "Not Aware at all" of MLCS software packages within their business. The next survey question asked these SMEs if their

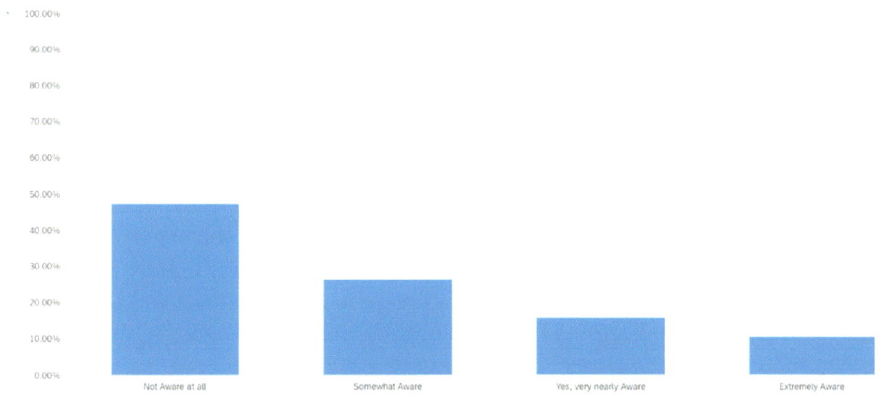

Figure 3.
Awareness of MLCS software packages within the business.

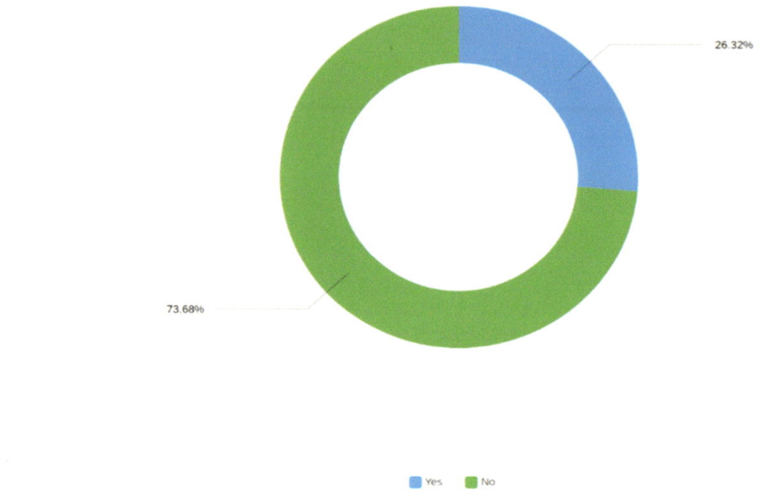

Figure 4.
SMEs recognizing existing software having ML to detect cyber attacks.

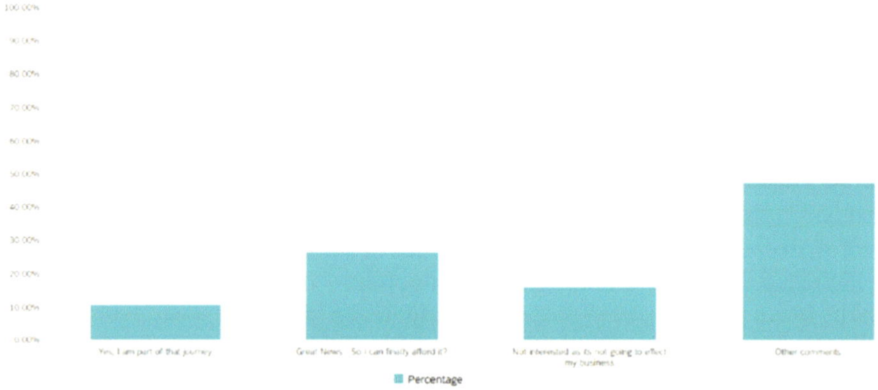

Figure 5.
SMEs thoughts on AI and ML usage and its affordability.

existing cyber security software packages supported ML to detect cyber-attacks as shown in **Figure 4**.

Figure 4 showed that 73.7% of SMEs said they were aware of their cyber security software packages supported ML to detect cyber-attacks. Out of those who said "No" were 26.3% of the respondents. The SMEs reasons were "due to costs" and the fact that they "Did not think that ML will be effective in cyber security and hence did not use or need this feature". The survey then asked SMEs if they knew of the recent Intuit research study in collaboration with the 2019 Gartner CIO survey that showed findings of SMEs already adopting next-generation AI and ML technology for many parts of their business already to combat the increase in cyber-attacks and that these costs were now affordable. **Figure 5** below visualized the results showing that 10% said they were "part of the journey" in attaining this intelligent software and 25% of the respondents were shocked that they could finally afford it.

Figure 5 showed 15% of SMEs saying they were "Not interested as it's not going to affect their business". A large percentage of 47% had other comments such as "having no idea about ML software", "Only just become aware that Defender used machine learning" and some saying, "Never heard about it" and some "wanting to need time to understand the benefits". The survey pointed out the average ransomware attack payout was now $177,000 according to Gartner 2020. Based on the fact explained, the survey went on to ask SMEs about their strategic planning in adopting ML to implement within business. **Table 1** showed 42% of SMEs were already talking to experts in this field, whilst 16% would rather concentrate on sales and marketing than cyber security, and still debating the perspective of costs.

SMEs thoughts on ML adaptation for cyber security	Percentage
None—I would rather just concentrate on the budget for Sales and Marketing than Cyber Security	16
Yes—I am trying to figure this out from a cost perspective	16
Yes—already talking to experts in this field	42
Any other comments to add	26

Table 1.
SMEs thoughts on ML adaptation for cyber security.

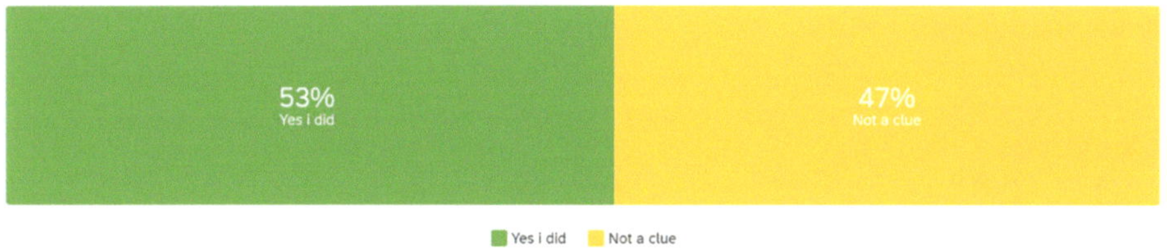

Figure 6.
SMEs knowledge of MLCS in predicting malware.

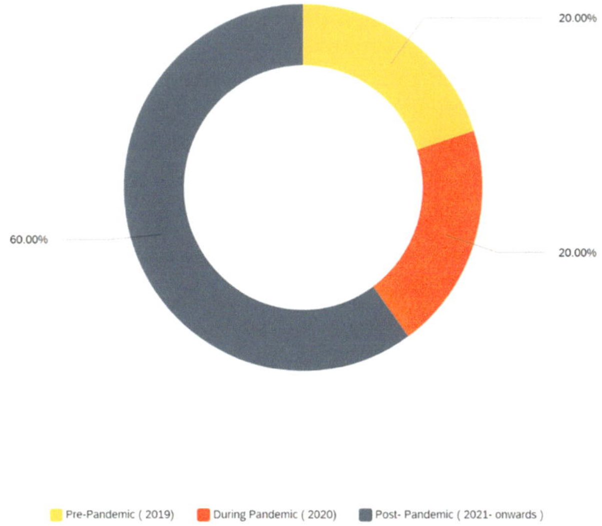

Figure 7.
SMEs being victims of cyber attacks.

Interestingly 26% of SMEs had varied strategic plans such as "*Cloud back up*", "*advice taken from Cyber Resilience Centre for Wales*", "*Investing in training to adopt a robust ISMS that will be conform with ISO 27001*", "*Make backups*" and "*Ransomware is something we deal with as with any other business outage*". The survey continued to ask SMEs if they knew MLCS software packages (compared to regular detection software) are trained in detecting a range of malware and can proactively detect, foresee breaches, and predict the types of malwares that might infiltrate the network.

Figure 6 showed that 53% said "Yes" to the statement above and 47% of SMEs did not have a clue.

The survey then asked the question to SMEs if their business was a victim of cyber-attacks (pre and post-pandemic). **Figure 7** below shows that 20% of SMEs surveyed were victims of cyber-attacks pre-pandemic (2019) and during the pandemic (2020).

This then left 60% of SMEs being victims of cyber-attacks post-pandemic (2021-onwards). One SME victim in the case study said, "*We had not secured our server with the latest updates. We took the server offline and had most of our files available on our computers still so lost a small portion. No customer data held on the server.*" Another SME victim said that "*We employed a Social Media company who sadly clicked an email message which got our Instagram account hacked. We lost 30,000 followers. That's it really, we no longer use them and have learnt while hacking is not nice, it happens and the companies you bring in to help need to be registered to confirm they understand the damages they can cause to small business by not following a simple set of do's & don'ts.*"

5. Conclusion

Any organization, regardless of the size, could be susceptible to malware threats and risk becoming victims of cybercrime, unless proper defense mechanisms are put in place. The failure to do so will pose threats to personal privacy and national security. The threats posed by malware are not a new phenomenon for large organizations in comparison to that SMEs. Also, large companies are equipped with resources both in terms of financial and technical capabilities, with policy defense mechanisms in place, whereas SMEs have limited resources, face budget constraints, and often lack in-house technical skills. That makes SMEs an attractive target for cyber-criminals, and breaches of personal data will have immediate consequences on the organizations and will impact the company finances, corporate image, and trustworthiness of the organization. Therefore, the onus is on the large organization to support SMEs with the provision of technical and policy know-how to protect SMEs from malware attacks.

The overarching focus must be to understand the different variants of malware, act on them promptly, manage the impact of the data breaches ensure the security of personal data, avoid recurrence by adherence to guidelines and regulations set out in the GDPR, to protect the interests of the data subjects, employees, the organization, and not forgetting the impact of bad publicity on the organization itself regardless of the size or the brand name of the enterprise.

Whilst the research is strong on the benefits of emerging technologies such as AI and ML in the detection of malware and intelligent cyber-attacks, MLCS software packages do take on an important role in the SME ecosystem in combatting these threats. Collaboration between technology, organization, and the understanding of human awareness, might still need some catching up to do. Through various methods of engagement within the industry coupled with education and training of the workforce, regularly conducted SME surveys such as the one in this paper, will give better meaning and understanding to the structure of how SMEs are having to cope and keep up with technology evolving in keeping their data safe and secure. Whilst AL and ML are emerging technologies that are maturing in their capabilities to protect intelligent cyber warfare, life in general for SMEs at the moment is trying to keep the balance between running their business running and prioritizing financially between sales, marketing, and now cyber. Support from decision makers and business owners taking on this responsibility will still be a long and windy road, where technology will finally try and meet the demands and affordability of the uses of MLCS within the SME business. Allowing for the costs to become affordable will rely on the supply and demand of the supply chain driving the affordability economically for all SMEs to get on the bandwagon of using intelligent software without breaking the bank when trading online to continuously keep their data safe and secure in line with GDPR within the context of the UK.

The novelty is to apply metaheuristic algorithms for detecting and predicting malware in SMEs and to optimize the problem. Also, more investigation of the likely malware is proposed to optimize security in SMEs.

6. Recommendations

All the software applications used must have integrated protection by design and default. That is an important precautionary measure to avoid incidents of breaches and to mitigate risk vulnerabilities associated with the security of sensitive data thus safeguarding the rights of the data subject. Also, risk impact assessment must be undertaken to evaluate the susceptibility to risks and to take measures to mitigate the risks.

The implementation of a consolidated uniform global level data protection mechanism is an imperative requirement to protect user information collected by small, medium, and large enterprises, and meaningful, robust provisions must be embedded in it to bring prosecutions against the perpetrators.

Within the SME market, decision-makers play an important role in keeping their business safe. With the onset of GDPR, the role of the data controller should allow for better understanding and safekeeping of the types of protection the SME business will use in order to keep their data safe. Having the right technology in place does have its advantages, however, whilst many SMEs still play the game of having "no protection", many still use traditional methods that are getting out of date. With the intelligence of the underworld growing, AI and ML software are becoming highly sought after at a cost to combat these threats. SME organizations are being advised to keep their level of security higher and use the right technology to help. This balance in getting the right IT expertise, coupled with the right costs of products, will inevitably help SMEs protect and secure their data. SMEs will surely benefit from using more intelligent controlled environments with applied machine learning techniques thus this demand will hopefully favor on costs. That said, the human engagement within this cycle is room for improvement, and the need to be trained and awareness raised in the workforce. This still remains a gap that will need to be managed by SMEs to stay ahead of the cyber warfare upon us already!

Acknowledgements

This paper has been supported by the British Council, KESS2—Knowledge Economy Skills Scholarships, Cardiff School of Technologies—Cardiff Metropolitan University, Cyber Resilience Wales, and Aytel Systems Ltd., Cardiff, UK.

Conflict of interest

The authors declare no conflict of interest.

Author details

Nisha Rawindaran[1], Liqaa Nawaf[1*], Vibhushinie Bentotahewa[1], Edmond Prakash[2], Ambikesh Jayal[3], Chaminda Hewage[1] and Daniyal Mohammed N. Alghazzawi[4]

1 Cardiff Metropolitan University, Cardiff, UK

2 University of Creative Arts, Farnham, Surrey, UK

3 University of Canberra, Bruce, Australia

4 King Abdulaziz University, Jeddah, Saudi Arabia

*Address all correspondence to: lllnawaf@cardiffmet.ac.uk

References

[1] European Union Agency for Cyber Security. Phishing Most Common Cyber Incident Faced by SMEs. 2021. Available from: https://www.enisa.europa.eu/news/enisa-news/phishing-most-common-cyber-incidents-faced-by-smes

[2] Huq S. Ransomware: the number one cyber threat for enterprises and SMEs. 2022. Available from: https://www.ncsc.gov.uk/blog-post/ransomware-the-number-one-cyber-threat-for-enterprises-and-sme

[3] Daniel K, Andreas J. Evaluation of AI-based use cases for enhancing the cyber security defense of small and medium-sized companies (SMEs). Electronic Imaging. 2022;**34**:1-8

[4] Pohlmann N. Cyber Security. The Textbook for Concepts, Principles, Mechanisms, Architectures, and Properties of Cyber Security Systems in Digitalization (transl. from german). 2019. Available from: https://doi.org/10.1007/978-3-658-25398-1_15. [Accessed: August 26, 2021]

[5] Rawindaran N, Jayal A, Prakash E, Hewage C. Cost benefits of using machine learning features in NIDS for cyber security in UK small medium enterprises (SME). Future Internet. 2021;**13**:186. DOI: 10.3390/fi13080186

[6] Valdetero J. Do All Malware Attacks Need to be Reported under the GDPR?. 2021. Available from: https://www.gtlaw-dataprivacydish.com/2021/02/do-all-malware-attacks-need-to-be-reported-under-the-gdpr/

[7] Towergate. Cyber Attacks and Security Threats—The Impacts of Cyber Attacks and How SMES Can Help Prevent Them. 2020. Available from: https://www.towergateinsurance.co.uk/liability-insurance/smes-and-cyber-attacks

[8] NortonLifeLock Employee. What is a Trojan?. (N.D). Available from: https://uk.norton.com/internetsecurity-malware-what-is-a-trojan.html

[9] Get Support IT Services. What Is a Trojan Horse? The Essential Guide for Small Business. 2020. Available from: https://www.getsupport.co.uk/blog/2020-12/what-is-a-trojan-horse-the-essential-guide-for-small-business/

[10] Kaspersky. What is a Trojan Horse and What Damage Can It Do?. (N.D). Available from: https://www.kaspersky.co.uk/resource-center/threats/trojans

[11] Zainab A et al. Phishing attacks: A recent comprehensive study and a new anatomy. Frontiers in Computer Science. 2021;**3**:563060. DOI: 10.3389/fcomp.2021.563060

[12] Checkpoint. The 5 Most Expensive Phishing Scams of all Time. (N.D). Available from: https://www.checkpoint.com/cyber-hub/threat-prevention/what-is-phishing/the-top-5-phishing-scams-of-all-times/

[13] Kaspersky. Ransomware Attacks and Types—How Encryption Trojans Differ. (N.D). Available from: https://www.kaspersky.co.uk/resource-center/threats/ransomware-attacks-and-types

[14] Pavilion. The Biggest Ransomware Attacks of 2021. 2021. Available from: https://www.pav.co.uk/blog/the-biggest-ransomware-attacks-of-2021/

[15] Drapkin A. 82% of ransomware attacks target small businesses, Report

Reveals. 2022. Available from: https://tech.co/news/82-of-ransomware-attacks-target-small-businesses-report-reveals#:~:text=twitter-,82%25%20of%20Ransomware%20Attacks%20Target%20Small%20Businesses%2C%20Report%20Reveals,employees%20are%20most%20at%20risk.&text=Small%20businesses%20are%20increasingly%20targeted,by%20ransomware%20recovery%20specialists%20Coveware

[16] ICO. Ransomware and Data Protection Compliance. (N.D). Available from: https://ico.org.uk/for-organisations/guide-to-data-protection/guide-to-the-general-data-protection-regulation-gdpr/security/ransomware-and-data-protection-compliance/

[17] Kaspersky. What is Spyware? (N.D). Available from: https://www.kaspersky.co.uk/resource-center/threats/spyware

[18] Phillips G. How to Protect Yourself From Unethical or Illegal Spying. 2019. Available from: https://www.makeuseof.com/tag/how-to-protect-yourself-from-unethical-or-illegal-spying/

[19] Ahaskar A. Spyware: How They Impact Enterprises and How to Spot an Infection. 2021. Available from: https://www.spiceworks.com/it-security/cyber-risk-management/articles/spyware-threat-against-enterprises/

[20] Malwarebytes. Adware. (N.D). Avaailable from: https://www.malwarebytes.com/adware

[21] ICO. Who Needs to Document Their Processing Activities?. (N.D). Available from: https://ico.org.uk/for-organisations/guide-to-data-protection/guide-to-the-general-data-protection-regulation-gdpr/documentation/who-needs-to-document-their-processing-activities/

[22] European Commission. Do the Rules Apply to SMEs?. (N.D). Available from: https://ec.europa.eu/info/law/law-topic/data-protection/reform/rules-business-and-organisations/application-regulation/do-rules-apply-smes_en

[23] PrivacyPerfect. GDPR for SMEs: Benefit or Burden?. 2019. Available from: https://blog.privacyperfect.com/gdpr-for-smes-key-points

[24] Intersoft Consulting. Art. 33 GDPR-Notification of a Personal Data Breach to the Supervisory Authority. (N.D). Available from: https://gdpr-info.eu/art-33-gdpr/

[25] GOV.UK. Cyber Security Breaches Survey 2022. (N.D). Available from: https://www.gov.uk/government/statistics/cyber-security-breaches-survey-2022

[26] Tirumala SS, Nepal N, Ray SK. Raspberry pi-based intelligent cyber defense systems for SMEs and smart-homes: An exploratory study. EAI Endorsed Transactions on Smart Cities. 2022;**6**(18):e4-e4

[27] Kshetri N. Economics of supply chain cyberattacks. IT Professional. 2022;**24**(3):96-100

[28] The Guardian. NHS Ransomware Attack: What Happened and How Bad is it?. 2022. Available from: https://www.theguardian.com/technology/2022/aug/11/nhs-ransomware-attack-what-happened-and-how-bad-is-it. [Accessed: August 25, 2022]

[29] Cruzado CF, Rodriguez-Baca LS, Huanca-López LG, Acuña-Salinas EI. Reference framework "HOGO" for cybersecurity in SMEs based on ISO 27002 and 27032. In: 2022 12th International Conference on Cloud

Computing, Data Science & Engineering (Confluence). IEEE Xplore Digital Library; 2022. pp. 35-40

[30] Ugli MIB. The importance of data mining In retail industry. International Journal of Progressive Sciences and Technologies. 2021;**28**(1):216-223

[31] Sayad L, Bouallouche-Medjkoune L, Aissani D. An electromagnetism-like mechanism algorithm for the router node placement in wireless mesh networks. Soft Computing. 2019;**23**(12):4407-4419. DOI: 10.1007/s00500-018-3096-y

[32] Gruschka N, Mavroeidis V, Vishi K, Jensen M. Privacy issues and data protection in big data: a case study analysis under GDPR. In: 2018 IEEE International Conference on Big Data (Big Data). IEEE Xplore; 2018. pp. 5027-5033

[33] New Technologies and Challenges for Personal Identity—The Digital Health Society. Feb 2021. Available from: https://thedigitalhealthsociety.com/new-technologies-and-challenges-for-personal-identity/. [Accessed: August 5, 2022]

[34] Nawaf L. Optimizing IoT Security by Implementing Artificial Intelligence—Infosecurity Magazine. May 2020. Available from: https://www.infosecurity-magazine.com/next-gen-infosec/optimizing-iot-ai/. [Accessed: August 1, 2022]

[35] Nawaf LF, Allen SM, Rana O. Optimizing infrastructure placement in wireless mesh networks using NSGA-II. In: 20th International Conference on High Performance Computing and Communications, 16th International Conference on Smart City and 4th International Conference on Data Science and Systems, HPCC/SmartCity/DSS 2018. New York, NY, USA: IEEE; 2019

Chapter 4

Malware: Detection and Defense

Iyas Alodat

Abstract

In today's cyber security landscape, companies are facing increasing pressure to protect their data and systems from malicious attackers. As a result, there has been a significant rise in the number of security solutions that can identify malware. But how do you know if an image file is infected with malware? How can you prevent it from running? This blog post covers everything you need to know about malware in your images and how to prevent them from running. The malware will allow the attacker or un-legitimate user to enter the system without being recognized as a valid user. In this paper, we will look at how malware can hide within images and transfer between computers in the background of any system. In addition, we will describe how deep transfer learning can detect malware hidden beneath images in this paper. In addition, we will compare multiple kernel models for detecting malicious images. We also highly suggest which model should be used by the system for detecting malware.

Keywords: steganography, cyber security, malware, deep transfer learning, keras

1. Introduction

What is a Malware? Malware is an umbrella term that refers to anything that is malicious. It can be used to refer to malicious hackers, viruses, or even worms. Malware is a very broad term that refers to any malicious software. Annoying popups, spyware, viruses, ransomware, and worms are all examples of malware.

Identifying the Types of Malware, There are many different types of malware, depending on what the purpose of the attack is. Viruses are malicious software that can replicate themselves and infect files on your computer. Trojans are malicious software that masquerades as something legitimate, like a helpful PDF reader, but actually do something harmful. Worms are software that spreads across networks and computers, like the dreaded WannaCry attack. Spyware collects information about you and your computer's activity without your knowledge. Ransomware can lock your computer or your files until you pay a ransom. Malware can do many different things, but you can protect yourself by keeping your computer clean and being careful about what you download.

How Malware Gets into Image Files, Malware is most commonly found in image files online. The most common cases of this happening are with stock photos from websites like Shutterstock and iStock. Sometimes, malicious software is also embedded in a company's digital images. This malware can do anything from collecting user information to carrying out a denial-of-service (DoS) attack on your company's servers. Websites that include images in their content often receive images from a stock photo website. It's possible that the image may include malicious software. If the website does not have an image scanning system in place, malicious software could make its way onto your website without you even knowing.

The crime-as-a-service field is rapidly evolving, making innovative operating emerging trends available to cybercriminals in order for them to successfully achieve their goals. These technologies have evolved into cyber threats that could be suited to the cryptographic protocols used by consumers and businesses to combat cybercrime. One of the major difficulties is undoubtedly malware classification.

Malware that appears "attractive" today will be obsolete tomorrow, filled by others with wholly distinct or improved features [1]. And all the while, newer isolates of malware collaborate with older varieties. As a result, designation in the malicious cyber environment is highly complex [2, 3].

In such a context, traditional cyber risk intelligence rollbacks and indices (IOCs) are insufficient to combat the threat. Who alters his behavior after learning that he has been identified? The biggest strength of cybersecurity deep learning is its capability to benefit from this evolution in real-time and generate classification criteria without the need for human intervention. This allows us to determine whether a person is communicating with their workstation or an automaton in real-time. Or if a cyber criminal is attempting to steal or interact with a user profile from any part of the community (remote access to a Trojan horse).

Many Facebook users revealed when they inspected a partial shot for hidden photograph tags attached to users' photos that images can undertake a lot of data that is typically inaccessible to the human eye. The type of data linked with Instagram and Facebook pictures and photos is not significant compared to the complex approaches used by targeted attacks to create images that can convey malicious code or embezzle user data. In the past several years, there has been a significant increase in malware advertisements in the wild that use the new technique of data encryption to embed subliminal meaning in photos and files.

2. Information's concealment

Steganography can hide code in plain text, such as inside an image file. This means that messages or information can be hidden inside a nonconfidential text as a carrier of these messages and information. In this way, malicious parties use this technology to compromise devices by hosting an image on a website or sending a picture in an e-mail. Hidden data or a carrier file does not have to be images; in fact that digital images are just streams of bytes like any other file making them an especially effective way to hide secret text and other data [4].

The science of steganography is a form of obfuscation that is very different from cryptography, which is the practice of writing encrypted or encrypted messages. The encrypted messages are clearly hiding something: and require specialized decryption methods.

Steganographic letters are similar to conventional letters, but they cleverly hide something unexpected. For instance, consider the following statement: (He eats solely like Lucifer, what other rogue enjoys Durian!). We can determine the core notion behind how to obscure information by reading this communication using a known technique. The hidden message, "Hello, world," is not encrypted; the reader only needs to know how to interpret the message in a specific way to identify it, and we did not just have to add any extra data to the "carrier" in order to deliver it. Although the technique of hiding the data is far more complex, it is essentially the same concept on a reduced scale.

The mind is interpreting the secret message in plain language in the preceding example. However, computer algorithms read bytes rather than natural words. It turns out that this allows you to hide communications in plain text that are simple for algorithms to perform and evaluate while being nearly hard for humans to uncover without assistance. Indeed, due to the nature of photographic file formats, it is feasible to conceal not just text strings but also complete files in .jpg format and other image formats depending on the technology employed; this may also be accomplished without increasing the overall file size for the original image.

3. Create malware images

We will use photos from both benign and harmful archives to identify photographs using a deep learning system. We will only do a binary characterization (malware and benign class). This method can also be used to achieve multiclass grouping, assuming that each variant of malware file has images that are distinct from the others. If our dataset is complete, we convert all files to 256 × 256 image pixels (every pixel has a value ranging from 0 and 255) by following the procedures below for each image: First, read 8 bits from the file at a time. Second, treat the eight bits as a binary form and translate it to an integer. Third, enter the pixel value as a number.

A file with a maximum size of 64 KB will fit a 256 x 256 image. For any file with a size greater than 64 KB, the remaining contents would be dropped. On the other hand, if the file size is smaller than 64 KB, the remaining image would be padded with 0's. Since the identification of malware is performed in real-time, we need to identify the picture as benign or malware within seconds. Keeping the image generation process quick and fast would help us save precious time.

4. Steganography hides information

Take a look at a few of the most basic methods for hiding text in a digital image. One straightforward method is to simply insert the material into the file at the end. These works do not prohibit the photograph from being displayed regularly, nor do they alter its esthetic look. We merely put "hello world" to the file's conclusion. The hex dump output shows us putting the extra bytes.

A program can easily read or discard the plain text string. In this scenario, we'll invert the hexadecimal number and output it in plain English using a software. For example, a received image displays a picture in a photograph viewer application ordinarily, but when examined with the WinRaR archiving utility, we can find that the unpacked.jpg file contains a concealed 28-byte text file.

These types of basic approaches can be helpful in collecting user data, but they do have some disadvantages. First, they inflate the file size, and second, they change the file hash. It's still very convenient for security tools to spot because of its unexpected format.

The best way is to enter into the code at the binary stage and deal with the least important bits (LSBs) of each pixel. Pixels can be represented in a 3-byte color image, one per RGB each (red, green, blue). Suppose we have three bytes

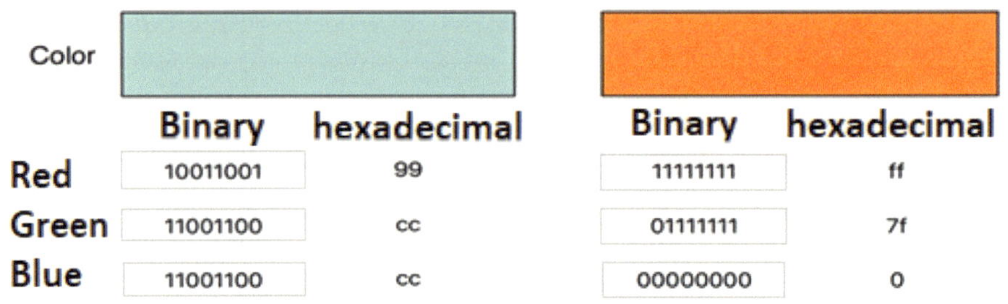

Figure 1.
Color with binary to swap the least important bits (LSBs).

representing a particular color as seen in **Figure 1**. You should swap the last four bits of the orange code with the first four bits of the turquoise code, to produce the composite RGB [5].

If we write software to read and extract these last four bits separately, we have effectively hidden the purple signal within the orange color software. Two pixels for the price of one, as there is no increase in file size. We can send our cryptic information without doubling the bandwidth of the actual message or modifying the file format, thus simple detection approaches that rely on examining files to find them are rendered obsolete. In actuality, the code is extremely cluttered before the attacker reassembles it.

In a sense, this ensures that an intrusion will utilize the last four bits of encoding RGB values to write extra information without interfering with the image's graphic display or increasing the file size. Another program will then read the hidden data and use it to reassemble a malicious script or sort customer data.

LSB processing is one of several steganography processes. There are numerous other instances in which photographs and other file formats might be modified to conceal a concealed message. To relay secret communications, the attackers employed information buried in network protocols, a technique known as "network hiding." The approach is the same in both cases: hide in clear view by downloading an invisible message to the accessible carrier.

Steganography for shielding information has affected both Windows and Mac OS operating systems. An intruder has been found to use cryptography to conceal portions of the ransomware attack code, add malicious JavaScript, and even download encryption software.

5. How to detect malware in an image file

In order to detect the presence of malware in an image file, you must first understand the types of malware that can be embedded into an image file. The different types of malware you may find in an image file include: There are many ways malware can hide within an image file, but luckily there are also several ways to detect it. At the network level, malware detection can be done through an antivirus program. Some administrators prefer to use a signature-based antivirus program because it can detect known viruses that are already in the wild. Signature-based detection is particularly helpful against viruses that are polymorphic, meaning they can constantly change their code and evade detection. Other malware may be detected by a heuristic method, which means the antivirus program looks for suspicious activity. This method can catch new viruses before they are added to the antiviruses' signature database.

6. SHA and MD5 checksum

If you are working with a file that has a large file size, SHA or MD5 checksum can be very useful in checking for malware. Many online tools can do this for you, or you can use a hex editor to calculate and compare the checksum for the file in the image. You'll need to download the file from the source and check the file size, then download the file from your image and check the file size to compare the two. If there is a difference in the file size, you may want to investigate further. If you are working with a smaller file, it may be easier to use a hex editor to view and compare the file directly.

7. Virus and worm signature detection

When dealing with image files and applications, many different types of malicious software can be hidden inside them with little or no indication. It's important to recognize what some of these indicators are so you can protect your network and its users from any potential dangers. In image files, viruses and worms are often obfuscated and hidden inside data sections. Viruses can also be hidden in executables as a DLL, EXE, or other file types. Some viruses and worms, such as Ramen and MyDoom, can be detected by signature because they have been seen before, and antivirus software vendors have created signature definitions to detect them. You can check for DLLs, EXEs, and other executable files hidden in the data section of the image file using a hex editor. You can also use antivirus software to scan the image file and look for specific signatures.

8. Hex editors and PE viewers

Depending on the complexity of the malware, it may be difficult to detect in an image file. Viruses and worms can be difficult to detect, but you can use a hex editor or a PE viewer to check an image's data section. This can also be helpful when detecting malicious code in an image file. You can view the hex data of an image file by opening it in a hex editor. This will show you any data that's been added to the image file, such as hidden code. If you are working with an EXE file, you can use a PE viewer to see the data that has been added. You can also use a debugger to debug the application and find any malicious activity.

9. Experiments setup simulation and dataset

The dataset was dealt with by Hacettepe University's Computer Engineering Multimedia Information Lab. It provides an RGB-based Core Fact Dataset for evaluating vision-based multiclass malware identification studies [6]. We used Keras Framework with TensorFlow in back-end, this about for deep learning libraries. For manipulate and process our images, we use Pandas and Scikit-leaning libraries. All experiments ran using Python 3.7 with notebook IDE.

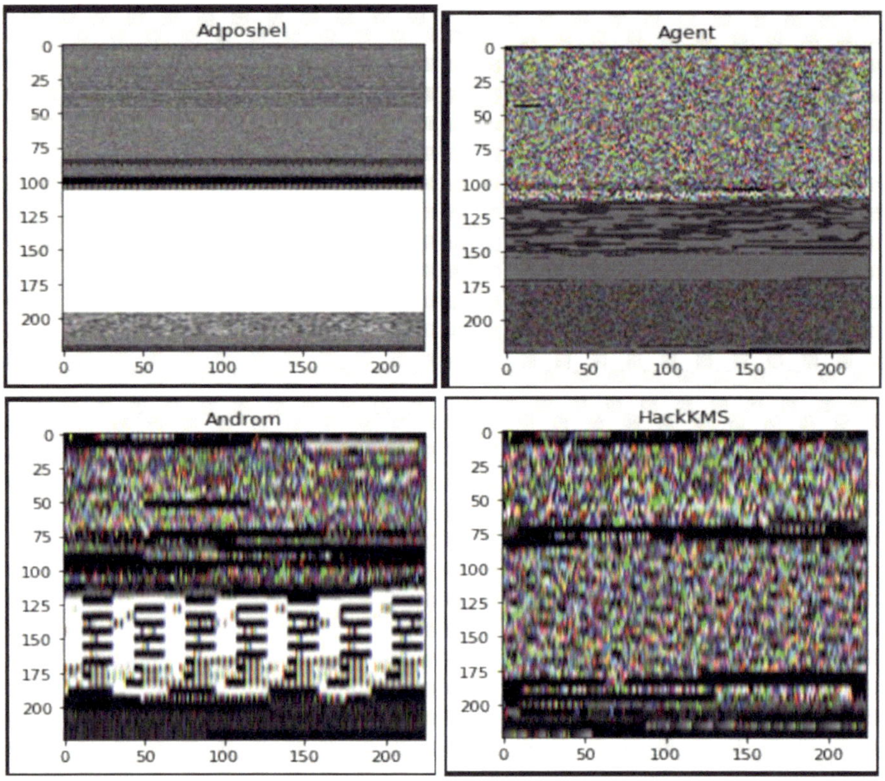

Figure 2.
Malware sample from dataset.

A Convolutional Neural Network (CNN, or ConvNet) is a sort of multilayer neural network designed to recognize visual patterns directly from pixel pictures with minimal pre-processing [7]. **Figure 2** shows an example of malware pictures. The ImageNet project is a massive graphic collection designed to be used in graphical object identification application testing. We utilized Keras, a deep-learning package. We simulate 18 types of malware using several deep transfer learning (DTL) models such as MobileNetV2, VGG16, and ResNet [8–12].

10. Performance evolution

We will utilize Classification Accuracy, Confusion Matrix, and ROC curve to assess the effectiveness of our systems.

Classification Precision It is a model classification metric in which the number of right predictions is compared to the total number of predictions produced by each model. Matrix of Confusion is one of the evaluation techniques that use the model's output and four categories: True positives are values that are supposed to be positive, whereas false positives are values that are expected to be positive but turned out negatively, making them fake. True negative values are those that are predicted to be negative and hence are true. False negative denotes Values that are predicted to be negative but are positive, hence they are false [13, 14].

DOI: http://dx.doi.org/10.5772/intechopen.108434

The ROC curve is a curve that compares two variables, the one being the genuine positive and the other being the false positive. True positive identifies the positive values that were accurately recognized by the model as positive. It defines false positives as negative values that were likewise considered to be negative by the classifier. The ROC curve graphic will show the true positive rate of a system in proportion to its false positive rate at various locations.

11. Results

We present our ideas to each Deep Transfer Learning using a confusion matrix, with each picture explaining a distinct form of malware. As we can see from the list

	Recall	Precision	Score
MobileNetV2	0.077187	0.076349	0.040323
InceptionV3	0.843847	0.834643	0.828221
ResNet50	0.810017	0.800643	0.808221
LittleVGG	0.768251	0.811926	0.775191

Table 1.
List of rates computed from a confusion matrix.

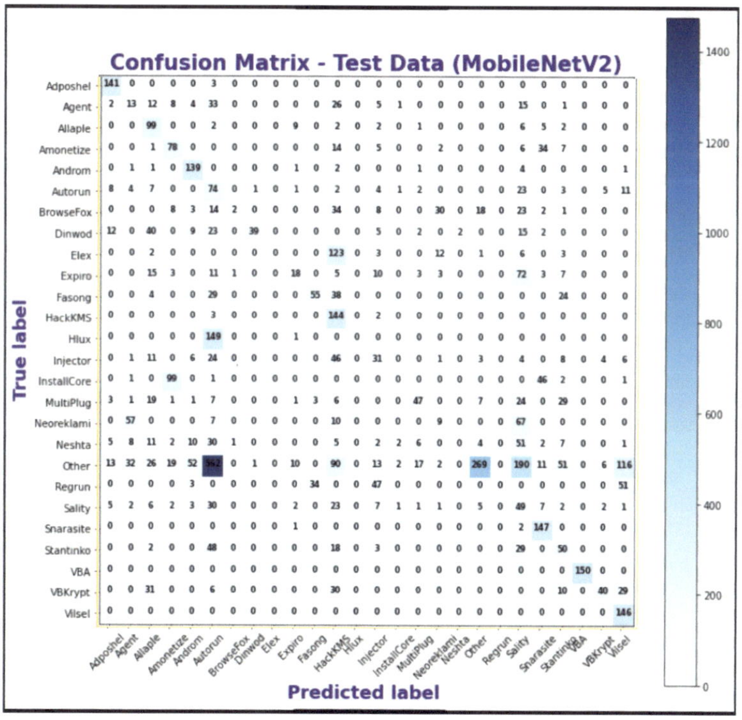

Figure 3.
Confusion matrix for mobileNetV2.

of rates in **Table 1**, we can determine that the best system for predicting our virus is MobileNetV2, which is the greatest pick from the list of rates.

In **Figure 3**, we can see from the confusion matrix that MobileNetV2 is the best model since it has a significant number of real values on the left and predicted values on the right. The proper productions will occur on the matrix's diagonal.

We have an emphasis on precision or specialization in our work, and these matrices can explain erroneous negatives. We require false negatives matrices for nonmalware detected by malware filters. In other words, positive class malware is malware, while false negatives are not malware. In this case, false negatives are preferable to false positives.

Table 2 shows the training, validation, and testing trials we conducted for each model. At this point, we may infer that one model detects malware far better than another. As the analysis from the confusion matrix shows, the best one is obviously MobileNetV2.

The ROC curve in **Figure 4** shows that it can accurately detect the kind of malware. The capacity to appropriately analyze and recognize a picture from another side whether it was typical.

	Training	Cross-val	Testing
MobileNetV2	0.94819713	0.94819713	0.95199621
InceptionV3	0.84456111	0.84809954	0.88111015
ResNet50	0.83881235	0.82547898	0.83109547
LittleVGG	0.90258367	0.89804627	0.92458974

Table 2.
Data accuracy after transfer learning—Performance metrics.

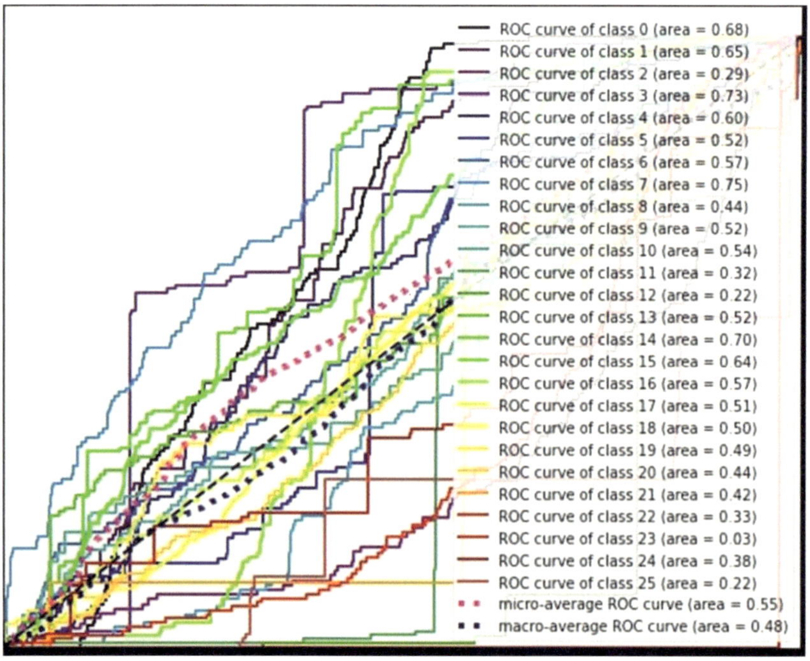

Figure 4.
ROC curve for mobileNetV2.

DOI: http://dx.doi.org/10.5772/intechopen.108434

12. Conclusion

Malware can infect any type of file, including images. Image file types that are most likely to be infected with malware are BMP, JPEG, PNG, and GIF. Additionally, malware can get into image files by being embedded in the image itself during the creation process. If you are using images in your marketing campaign or on your website, you need to make sure they are safe. Check your images to make sure they do not contain any malicious software. If you end up with an infected image, you could put your company's systems at risk. If you are unsure if an image is malicious, you can upload it to a website like VirusTotal. This service will scan the image for malware. This is the best way to make sure your images are safe.

Many photos are transmitted by e-mail and social media, which may be hiding a specific threat. Through the analysis that we have done in this study, we can reduce the dangers of malware hidden behind those images, using artificial intelligence techniques and machine learning, we were able to reduce these risks very well, and also reduce privacy violators by discovering these types of deceptive users through temptation and carrots that the hacker uses for this purpose.

Author details

Iyas Alodat
Jerash University, Jerash, Jordan

*Address all correspondence to: eyas.odat@jpu.edu.jo

References

[1] Dini G, Martinelli F, Saracino A, Sgandurra D. MADAM: A multi-level anomaly detector for android malware. In: International Conference on Mathematical Methods, Models, and Architectures for Computer Network Security. Berlin, Heidelberg: Springer; 2012. pp. 240-253

[2] Chandrasekhar AM, Raghuveer K. Intrusion detection technique by using k-means, fuzzy neural network and SVM classifiers. In: 2013 International Conference on Computer Communication and Informatics. 2013. pp. 1-7. DOI: 10.1109/ICCCI.2013.6466310

[3] Krizhevsky A, Sutskever I, Hinton GE. ImageNet classification with deep convolutional neural networks. In: Proceedings of the 25th International Conference on Neural Information Processing Systems - Volume 1 (NIPS'12). Red Hook, NY, USA: Curran Associates Inc.; 2012. pp. 1097-1105

[4] Al-Juaid NA, Gutub AA, Khan EA. Enhancing PC data security via combining RSA cryptography and video based steganography. Journal of Information Security and Cybercrimes Research. 2018;**1**(1):5-13

[5] Islam MR, Siddiqa A, Uddin MP, Mandal AK, Hossain MD. An efficient filtering based approach improving LSB image steganography using status bit along with AES cryptography. In: 2014 International Conference on Informatics, Electronics & Vision (ICIEV). 2014. pp. 1-6. DOI: 10.1109/ICIEV.2014.6850714

[6] Bozkir AS, Cankaya AO, Aydos M. Utilization and comparison of convolutional neural networks in malware recognition. In: 2019 27th Signal Processing and Communications Applications Conference (SIU). 2019. pp. 1-4. DOI: 10.1109/SIU.2019.8806511

[7] Xiao Y, Xing C, Zhang T, Zhao Z. An intrusion detection model based on feature reduction and convolutional neural networks. IEEE Access. 2019;7:42210-42219

[8] Ahmim A, Maglaras L, Ferrag MA, Derdour M, Janicke H. A novel hierarchical intrusion detection system based on decision tree and rules-based models. In: 2019 15th International Conference on Distributed Computing in Sensor Systems (DCOSS). 2019. pp. 228-233. DOI: 10.1109/DCOSS.2019.00059

[9] Vinayakumar R, Alazab M, Soman KP, Poornachandran P, AlNemrat A, Venkatraman S. Deep learning approach for intelligent intrusion detection system. IEEE Access. 2019;7:41525-41550

[10] Riyaz B, Ganapathy S. A deep learning approach for effective intrusion detection in wireless networks using CNN. Soft Computing. 2020;**24**(22):17265-17278. DOI: 10.1007/s00500-020-05017-0

[11] Yang L et al. A theory of transfer learning with applications to active learning. Machine Learning. 2012;**90**(2):161-189. DOI: 10.1007/s10994-012-5310-y

[12] Tan C, Sun F, Kong T, Zhang W, Yang C, Liu C. A survey on deep transfer learning. In: Kůrková V, Manolopoulos Y, Hammer B, Iliadis L, Maglogiannis I, editors. Artificial Neural Networks and Machine Learning – ICANN 2018. ICANN 2018. Lecture Notes in Computer Science. Vol. 11141. Cham: Springer; 2018. DOI: 10.1007/978-3-030-01424-7_27

[13] Alodat M, Abdullah I. Surveillance rapid detection of signs of traffic Services in Real Time. Journal of Telecommunication, Electronic and Computer Engineering (JTEC). 2018;**10**(2-4):193-196

[14] Alodat M. Predicting student final score using deep learning. In: Bhatia SK, Tiwari S, Ruidan S, Trivedi MC, Mishra KK, editors. Advances in Computer, Communication and Computational Sciences. Advances in Intelligent Systems and Computing. Vol. 1158. Singapore: Springer; 2021

Section 3

Ransomware

Chapter 5

Perspective Chapter: Ransomware

Arun Warikoo

Abstract

Ransomware refers to a type of malware that encrypts files on an infected computer and holds the key to decrypt the files until the victim pays a ransom. Ransomware has seen explosive growth over the past few years and has rapidly evolved into a highly lucrative business model. Sophisticated advanced persistent threats (APTs) are employing ransomware to maximize their profits with multiple layers of monetization strategies. New versions appear frequently with ever-evolving tactics and techniques making detection harder. In this chapter, we present a brief history of ransomware, top threat actors employing ransomware, tactics used, and key strategies firms need to deploy to prevent, detect, and respond to ransomware in attacks.

Keywords: ransomware, extortion, threat actor groups, tactics, prevention, detection, response

1. Introduction

Ransomware attacks have emerged as one of the most prominent cyberattacks in the last 5 years affecting organizations globally. The Verizon Data Breach Investigation Report (DBIR) 2021 states that 37% of global organizations said that they were hit by ransomware [1]. The world saw a 151% year-on-year increase in the number of ransomware attacks by mid-2021 [2].

Ransomware is a family of malware that is designed to block or limit victims from accessing their system by either locking the system's screen or encrypting files on a system until ransom is paid. Ransom operators demand the victim to pay the ransom in crypto, usually, bitcoin.

Ransomware variants are of two types—encryptors and lockers [3]. The encrypting ransomware encrypts the files on the victim's machine and demands a ransom for the decryption key. On the other hand, lockers do not encrypt the file but lock the victim's system so that the files are inaccessible.

Ransomware tactics and techniques have evolved considerably over the years. The evolution of ransomware can be broken down into three key timeframes: pre-2014, between 2015 and 2017, and post-2017. During the pre-2014 era, ransomware attacks were widespread but random with a very low ransom demand. Post-2015, attackers started deploying ransomware post-exploitation. This shift reduced the number of victims that an attacker could exploit, but this gave operators much more control over ransomware deployment. This enabled targeted and successful encryption of files on the victim's network and justified demands for a higher ransom. Led to the rise of targeted attacks that were highly successful leading to a higher ransom demand. Post-2017, the ransomware threat landscape witnessed the emergence of ransomware as a service (RaaS) and big game hunting (BGH). Big game hunting refers to when attackers leverage ransomware to target large and high-value organizations [4].

Subsequent sections highlight a brief history on ransomware, how ransomware is distributed, high-profile ransomware groups, and how to prevent, defend, and respond to ransomware attacks.

2. A brief history of ransomware

This section details a brief history on ransomware from its inception as a petty cybercriminal act into what is now a billion-dollar cyber-crime industry.

The first known ransomware attack occurred in 1989 and targeted the healthcare industry. An individual known as Joseph Popp, an AIDS researcher, carried out the attack by mailing 20,000 floppy disks to the WHO AIDS conference event attendees [5]. The attacker employed social engineering to trick the victims by claiming that the disks contained a questionnaire to determine an individual's risk of acquiring AIDS. However, the disk also contained a malware that was dubbed as the AIDS Trojan that encrypted files on the victim's machine and displayed a message demanding a payment of $189 to a P.O. box in Panama in exchange for access to their files [5]. AIDS Trojan used a simple symmetric encryptor to encrypt file names and a decryption key was soon available to decrypt them [6].

Figure 1 highlights the timeline on ransomware since the launch of AIDS Trojan in 1989 to Hive in 2022.

The first modern ransomware, GPCode, was launched in 2004 and infected systems via phishing emails [6]. GPCode also known as GPCoder used symmetric encryption to encrypt files and requested $20 for a decryption key [5]. The Year 2006 saw the launch of Archievus that employed strong encryption for the first time and used an advanced 1024-bit RSA encryption [5]. Reveton emerged in 2012 as a ransomware locker, a variant that displayed fraudulent law enforcement messages accusing victims of committing a crime. The attackers threatened victims with jail time if the ransom was not paid [5]. The year 2013 saw the emergence of a ransomware strain known as CryptoLocker that was delivered via phishing emails. CryptoLocker

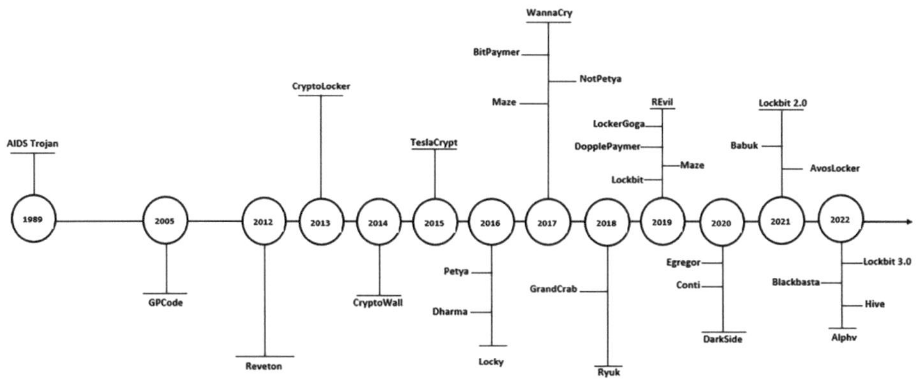

Figure 1.
Ransomware timeline.

used strong 2048 RSA encryption and was both a locker and a crypto variant [5]. CryptoWall gained notoriety after the downfall of the infamous CryptoLocker in 2014 and was widely distributed using various exploit kits and spam campaigns. TeslaCrypt gained notoriety in 2015 and targeted computer gamers. After a successful infection, the malicious program demands a $500 ransom for the decryption key; if the victim delays, the ransom doubles.

Petya emerged in 2016 as the first ransomware variant to not encrypt individual files but overwrite the master boot record and encrypt the master file table. These locked victims out of their entire hard drive more quickly than other ransomware techniques [5]. The infamous WannaCry shocked the world in 2017 and hit hundreds of thousands of machines across more than 150 countries. WannaCry spread via the Eternal Blue vulnerability, an exploit leaked from the National Security Agency [5]. A major cyberattack began targeting Ukraine in June 2017 using a new variant on Petya known as NotPetya. NotPetya soon spread and impacted organizations globally. In 2018, a sophisticated ransomware variant known as Ryuk was released and became one of the most successful ransomware campaigns of its time. Ryuk attacks were targeted, and ransom amounts associated with Ryuk typically range between 15 and 50 Bitcoins, or roughly between $100,000 and $500,000 [7]. REvil, also known as Sodinokibi, first appeared in April 2019 and immediately became immensely successful [8]. Another ransomware variant by the name Maze was discovered in 2019 and introduced the tactic of double extortion wherein data are exfiltrated before ransomware deployment. Shortly after Maze disbanded in 2020, the Egregor RaaS double extortion variant appeared. 2020 saw the emergence of a Conti and Darkside that were responsible for major cyber incidents globally. LockBit 2.0, a new variant of Lockbit with advanced capabilities appeared in 2021. LockBit 3.0, the current version, was discovered in June 2022 and has added a Big Bounty Program (BBP) to its arsenal [9]. The year 2022 saw the fall of a notorious ransomware group known as Conti and

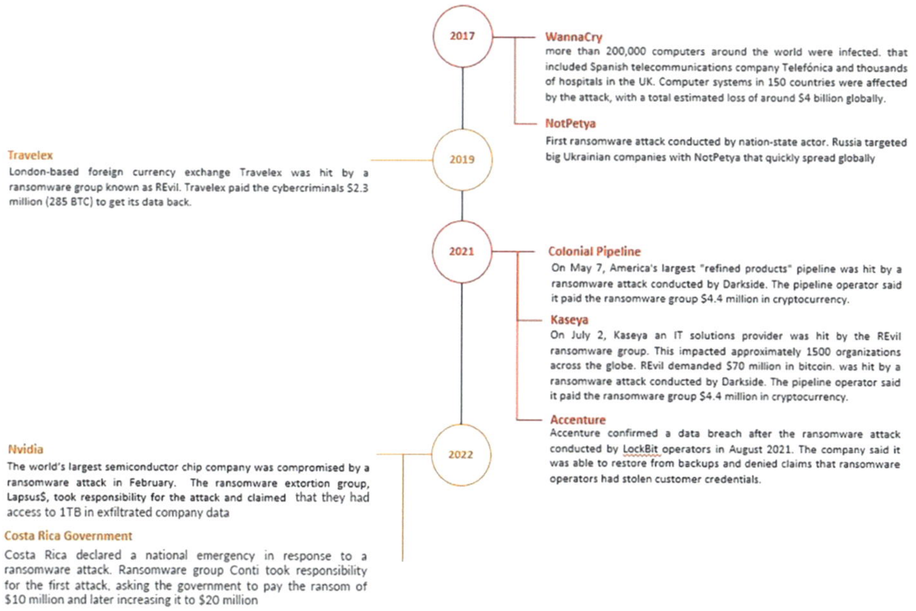

Figure 2.
Significant ransomware incidents.

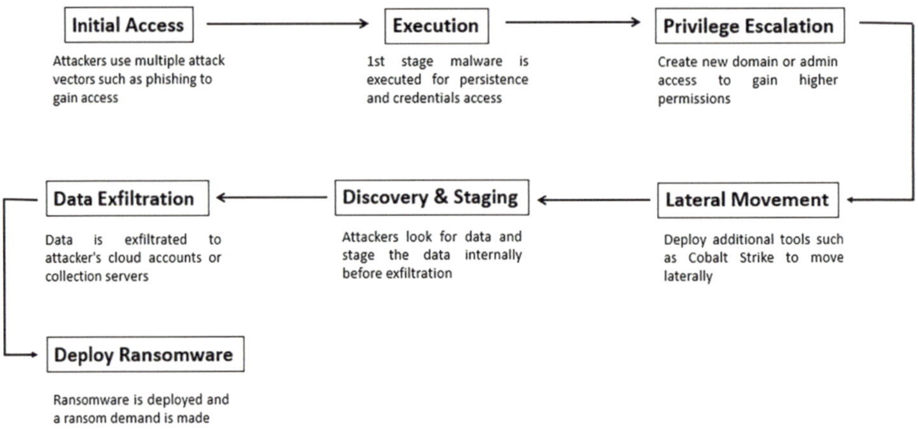

Figure 3.
Ransomware attack chain.

the emergence of new groups such as Blackbasta, Hive, and Quantum that continue to drive the ransomware threat landscape [10].

Figure 2 highlights significant ransomware incidents that have occurred over the last 5 years.

Figure 3 shows the various stages involved in a ransomware attack.

3. Distribution methods

Ransomware is spread through multiple distribution methods. These are as follows:

Phishing—The most common attack vector used by attackers as shown in **Figure 4**. Attackers send an email that is designed to lure the victim to open the weaponized office attachments. When the user opens the attachment (word or excel) and enables the macros, a malicious program is executed that executes a PowerShell command to download a 2nd stage malware from the Command and Control (C2) Server. Additional payloads are downloaded for lateral movement and once control is gained on the active directory

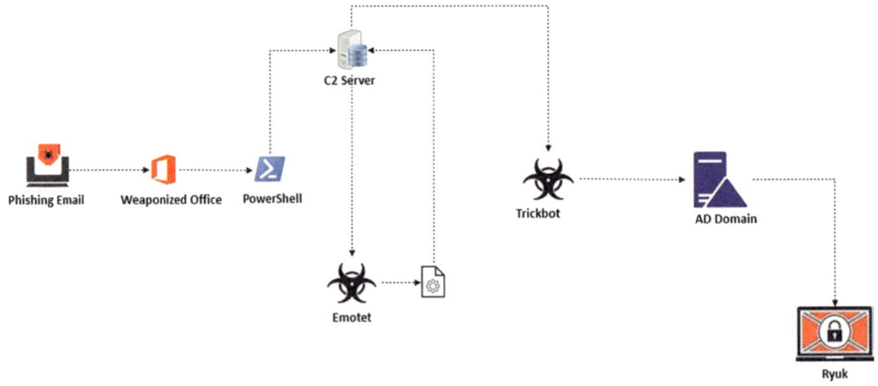

Figure 4.
Attack vector phishing.

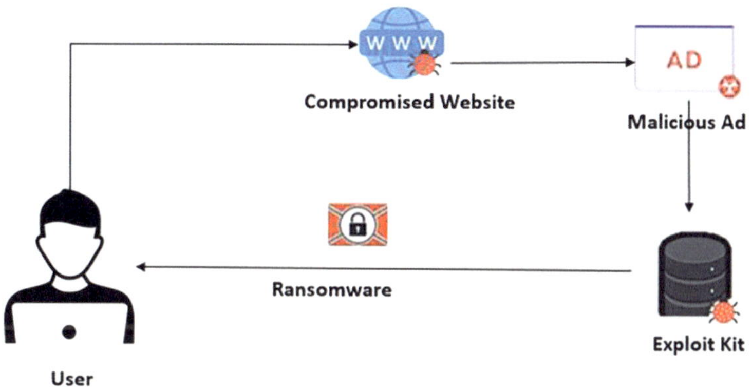

Figure 5.
Attack vector exploit kit.

(AD) domain, the attacker downloads ransomware as a final payload and deploys it to multiple devices.

Exploit kits—An exploit kit is a toolkit designed to exploit vulnerabilities on victim's system while web browsing. When a user visits a compromised website, the victim is redirected to another landing page. The victim's machine is scanned for any browser-based vulnerabilities and malware is downloaded. Ransomware groups employ malvertising to redirect users to the attacker's website, exploit is executed that leads to the eventual deployment of ransomware (**Figure 5**).

Buying credentials from access brokers—Attackers buy credentials from initial access brokers (IABs) to gain initial access. Remote desktop protocol (RDP) is the most common credential used to achieve a foothold.

Exploiting vulnerabilities—Ransomware operators also gain initial access by exploiting vulnerabilities in Internet-facing applications.

3rd Party Vendor—Supply chain has become the latest attack vector that has led to ransomware deployment.

4. Ransomware threat landscape

The modern ransomware threat landscape is driven by the advent of the ransomware as a service (RaaS) business model and the adoption of multiple levels of extortion.

4.1 Ranvsomware as a service

Ransomware as a service (RaaS) is a business model launched by ransomware operators wherein the operators sell ransomware to their customers known as affiliates in exchange of a cut from the ransom. The affiliates launch the cyberattack against the victims whereas negotiations with the victim are managed by the operator.

RaaS has taken ransomware to a whole new plane and is one of the primary reasons why ransomware attacks have become so frequent. RaaS business model is a win-win situation for all the parties involved. A report by Crowdstrike, a cybersecurity firm, states that the ransomware revenues in 2020 were around $20 billion, up from $11.5 billion the previous year [11].

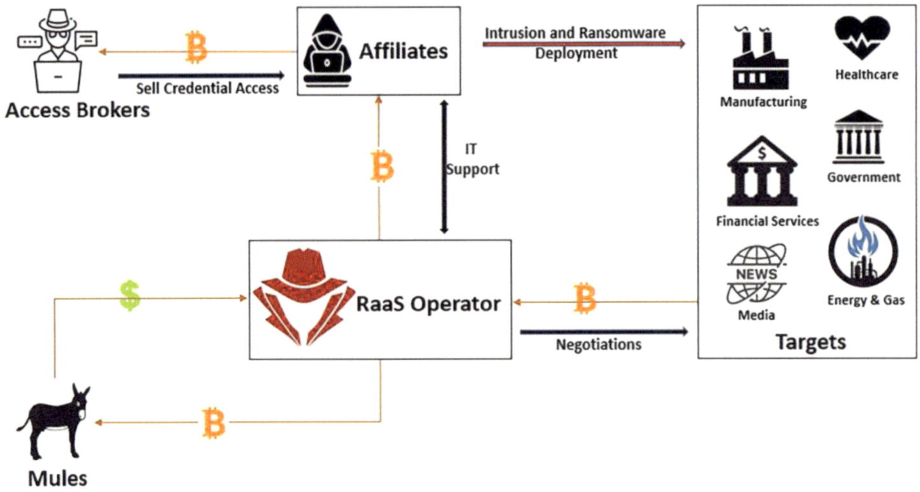

Figure 6.
The RaaS ecosystem.

The operator now only focuses on developing and monetizing its product. Less sophisticated actors with very little knowledge can enter the playing field, buy the service, and launch targeted attacks. **Figure 6** highlights the RaaS ecosystem. The ecosystem also comprises of initial access brokers (IABs) and Mules. Access brokers are an important component as they are the ones who scan the networks to look for vulnerabilities and gain credential access. Access brokers sell credentials access to the affiliates who leverage that for initial access during an intrusion. The RaaS operators handle the ransom negotiations with the victim. Mules complete the ecosystem and are used for converting cryptocurrency into real currency.

4.2 Extortion

The advent of extortion as a tactic has emerged as one of the major reasons for high-profile ransomware attacks in the last few years. Three are four types of extortion prevalent that are highlighted in **Figure 7**.

Single extortion refers to the deployment of ransomware post-exploitation. The attacker demands a ransom in exchange for decrypting the files.

Double extortion refers to attackers exfiltrating data before the deployment of ransomware. The attacker then threatens the victim to leak the data publicly. The Maze ransomware group pioneered this when they added double extortion as a tactic to their playbook. More threat actors followed suit had started to have dedicated leak sites (DLS) to release the stolen data.

In 2020, threat actors took extortion to another level and added DDoS attacks to encryption and data exposure threats. This is known as triple extortion. This was first performed by SunCrypt and RagnarLocker operators in the latter half of 2020 [12].

In 2021, a fourth level known as Quadruple extortion was introduced. With quadruple extortion, ransomware operators also reach out directly to a victim's customers and stakeholders, thereby adding more pressure to the victim. DarkSide operators employ the quadruple extortion scheme in some of their attacks by launching DDoS attacks and directly contacting customers through designated call centers [12].

Figure 7.
Types of ransomware extortion.

5. Ransomware groups

This section highlights some of the high-profile threat actors that have revolutionized ransomware campaigns.

Conti also known as Wizard Spider is a Russia-based cybercriminal operational since 2016 [13]. The group is known for being the operator of Ryuk and Conti ransomware variants and resorts to big game hunting (BGH). Conti used the Ryuk ransomware variant since September 2018 but switched to Conti in 2020 [14].

Carbon Spider also known as FIN7 is another Russia-based cybercriminal that operated since 2013. The group pivoted to ransomware and big game hunting in 2020 and marketed its own RaaS program dubbed as "DarkSide" [15]. In May 2021, the Colonial Pipeline ransomware attack made headlines across the globe that FBI attributed to the DarkSide group [16].

Pinchy Spider is a sophisticated cybercriminal operational since 2018 that is known to be operation of the REvil RaaS program [17]. Pinchy Spider is associated with some of the most high-profile ransomware attacks in history.

The LockBit group is a sophisticated cybercriminal operational since 2019. The group is known to consistently develop new tactics and techniques to stay ahead of other ransomware groups [12]. In 2021, a new variant known as Lockbit 2.0 was released that followed the RaaS model and LockBit 2.0 operators allegedly only work with experienced penetration testers [18].

Ransomware Name	Operated By	Operational Since	Written In	Encryption	Target Platforms	RaaS Model	Extortion	Big Bounty Program
TeslaCrypt	NA	2015	C++	AES256	Windows	No	No	No
WannaCry	Lazarus Group	2017	C/C++	RSA	Windows	No	No	No
BitPaymer	Indrik Spider	2017	C/C++	RC4 and RSA	Windows	No	No	No
DoppelPaymer	Doppel Spider	2019	C	RC4 and RSA	Windows	No	No	No
SunCrypt	SunCrypt	2019	GO, C & C++	RSA	Windows	Yes	Triple	No
Clop	FIN11 aka TA505	2019	C	RC4	Windows	Yes	Double	No
RagnarLocker	RagnarGroup	2019	C	Salsa20 RSA-1024	Windows		Double	No
Maze	Twisted Spider	2019	C	ChaCha20	Windows	Yes	Double	No
Egregor		2020	C/C++	ChaCha8 RSA-1024	Windows	Yes	Double	No
Ryuk	Wizard Spider aka Conli	2018	C/C++	AES-256 RSA-4096	Windows	Yes	Single	No
Conti		2020	C/C++	ChaCha18 RSA-1024	Windows Linux	Yes	Double	No
Darkside	Carbon Spider aka FIN7	2020	C	Salsa20 RSA-1024	Windows Linux	Yes	Double	No
Blackmatter		2021	C	Salsa20 RSA-1024	Windows Linux	Yes	Double	No
REvil aka Sodinokibi, GandCrab	Pinchy Spider	2019	C	RSA	Windows Linux	Yes	Triple	No
Lokbit	Lockbit Group	2019	C	AES-128	Windows	Yes	Double	No
Lockbit 2.0		2021	C	AES-128 Curve25519	Windows	Yes	Double	No
Lockbit3.0		2022	C	ChaCha256	Windows	Yes	Double	Yes

DOI: http://dx.doi.org/10.5772/intechopen.108433

Ransomware Name	Operated By	Operational Since	Written In	Encryption	Target Platforms	RaaS Model	Extortion	Big Bounty Program
HelloKilty	FiveHands Group	2021	C++	AES256 RSA2058	Windows	Yes	Double	No
AvosLocker	Avos Group	2021	C/C++	AES256	Windows Linux VMware ESX	Yes	Double	No
BlackCat aka Alphv	BlackCat group	2021	Rust	AES ChaCha20	Windows Linux VMware ESX	Yes	Triple	No
BlackBasta	BlackBasta group	2022	C++	ChaCha20 RSA-4096	Windows Linux VMware ESX	Yes	Double	No

Table 1.
Ransomware families.

BlackByte is ransomware as a service (RaaS) that first emerged in July 2021 and primarily exploits vulnerabilities to gain a foothold in the victim's environment [19].

Table 1 highlights significant ransomware families over the years with additional details such as operator, operating since, and encryption.

6. Ransomware prevention, detection, and response

Organizations need to take a multipronged approach to prevent and defend against a ransomware attack. The best strategy to tackle ransomware is a combination of prevention, detection, and recovery capabilities.

6.1 Prevention

Organizations need to have controls in place to cover all the distribution methods as highlighted in Section 3 and as part of defense-in-depth deploy controls at multiple levels.

At the network layer, organizations need to implement solutions such as Email Gateway Security and a sandbox solution to prevent against phishing campaigns which is the most common attack vector for ransomware. Web application firewalls (WAFs) enable in preventing initial access from exploits that target public-facing applications. Intrusion prevention systems (IPS) and content filtering solutions enable in preventing communication with command and control servers. Most sophisticated ransomware operations also exfiltrate data as a form of extortion. Data loss prevention (DLP) solutions are an important control for preventing against data leakage.

At the endpoint layer, apart from an anti-virus (AV) solution, organizations need to implement endpoint detection and response (EDR) solutions to detect malicious activity such as the spawning of a malicious process. In addition, organizations need to configure their information technology (IT) environment to prevent enabling of macros in documents received from outside the network without interrupting any business processes. It is also advisable to install browser protection and ad blocking on end-user workstations as this will prevent JavaScript-based malware from executing on the system [20].

Organizations must also have a robust vulnerability management program that focuses on hardening workstations and servers within the network. Attackers leverage exploit kits to exploit a vulnerability in systems and technologies. As an example, the Locky ransomware is frequently delivered via the Rig exploit kit that targets some of the Adobe flash vulnerabilities [20]. Therefore, it is imperative to have your systems and applications fully patched and up to date.

Most ransomware distribution methods require end-user interaction. Therefore, organizations need to create a robust security awareness training for their employees and train them on how attackers leverage social engineering to trick the users.

6.2 Detection

Detective controls play an important role in the fight against ransomware. In a modern ransomware attack, there is a significant dwell time before ransomware is deployed and executed. With the advent of big game hunting, attackers spend considerable time in identifying high-value targets post-compromise. The dwell can be as little as a few days and can go up to weeks before the deployment of ransomware. This enables defenders to deploy detective controls in order to identifying unusual activity pointing to an ongoing ransomware attack.

Ransomware detection can be done by mirroring the behavior depicted by various ransomware variants. Ransomware behavior involves the generation of network traffic to C2servers that includes domain name service (DNS) queries [21]. Detective controls such as an intrusion detection system (IDS) and a security information and event management (SIEM) have inbuilt signatures and rules to detect for such events. In addition, custom rules can be created to look for specific behaviors such as anomalous SMB traffic, creation of new privileged accounts, anomalous outbound traffic, and monitoring of processes and PowerShell. Alerts from detective controls are investigated in a Security Operations Centre (SOC). Alerts from detective controls have a high false positive rate and it is imperative that SOC analysts work with the threat detection team to tune the platforms and reduce the noise [22].

Organizations need to develop cyber fusion capabilities to tackle advanced persistent threats. This can be achieved by the creation of a Cyber Defense Centre (CDC) that comprises of teams such as CSIRT, Red Team, VA, threat detection, and cyber threat intelligence team (CTI). Cyber threat intelligence plays a vital role in providing intel on ongoing campaigns to ensure that your enterprise defenders are ready for the threat and know what to look for. Organizations also need to conduct Table Top exercises quarterly and look at specific scenarios based on intel received from the threat intelligence team.

6.3 Response

It is imperative to create a Ransomware Incident Response Plan that will be executed by an organization's computer security incident response team (CSIRT).

Identify the infected systems within the network and isolate the infected devices immediately. It is extremely important to determine the scope of the infection. Look for symptoms such as file name changes and service tickets from employees on not able to access files.

Secure your backup data by taking them offline and ensure that the backup data is not infected by running a full scan [23]. Restore compromised files with backup data once all the devices have been decrypted and running antivirus.

The incident response team must also identify the attack vector and chart out the attack timeline. This is important to help in identifying how the attack happened, identifying the control gaps, and preventing recurring ransomware attacks in the future.

Report the incident to law enforcement immediately as typical ransomware attacks involve data leaks. Within the United States, report the incident to the nearest FBI office which can help identify those responsible and prevent future attacks [23].

Once normal operations resume, it is always advisable and recommended to conduct a post-incident activity to review the lessons learned from the ransomware attack.

7. Conclusion

Ransomware has become one of the most prevalent cybercrimes that threat actors leverage to maximize their profits. Nation-state actors have also employed ransomware to maximize their geopolitical interests. Organizations can no longer rely on resiliency and backups to thwart such attacks. Threat actors have evolved their tactics and are now employing multiple layers of extortion to threaten victims. The advent of RaaS has made matters worse as less sophisticated attackers can also launch ransomware attacks by buying the service.

Organizations need to employ a multipronged strategy to prevent, defend, and respond to such persistent attacks. Firms need to invest deeply in building a cyber fusion model that focuses on developing collaboration and cohesiveness between various cyber defense teams. Cyber threat intelligence exchange with ISACs and law enforcement agencies is extremely important to gain an understanding on the campaigns of interest. It is also recommended to adopt a Zero Trust Approach while designing the network.

Nomenclature

AD	Active Directory
AIDS	Acquired Immunodeficiency Syndrome
APT	Advanced Persistent Threat
ATT & CK	Adversarial Tactics, Techniques, and Common Knowledge
AV	Anti-virus
BBP	Big Bounty Program
BGH	Big Game Hunting
BTC	Bitcoin
C2	Command and Control
CDC	Cyber Defense Centre
CISA	Cybersecurity and Infrastructure Agency
CSIRT	Computer Security Incident Response Team
CTI	Cyber Threat Intelligence
CVSS	Common Vulnerability Scoring System
DBIR	Data Beach Investigation Report
DLP	Data Loss Prevention
DLS	Data Leak Sites
DGA	Domain Generation Algorithm
DNS	Domain Name Service
DoS	Denial of Service
DDoS	Distributed Denial of Service
EDR	Endpoint Detection and Response
ECDSA	Elliptic Curve Digital Signature Algorithm
FBI	Federal Bureau of Investigation
IAB	Initial Access Broker
IT	Information Technology
IDS	Intrusion Detection System
IPS	Intrusion Prevention Systems
IR	Incident Response
ISAC	Information Sharing and Analysis Centers
MFA	Multi-Factor Authentication
NIST	National Institute of Standards and Technology
OSINT	Open Source Intelligence
RaaS	Ransomware as a Service
RDP	Remote Desktop Protocol
RSA	Rivest-Shamir-Adleman

SIEM	Security Information and Event Management
SOC	Security Operations Center
SQLi	Structured Query Language Injection
TB	Terabyte
TTP	Tactics Techniques and Procedures
VA	Vulnerability Assessment
VPN	Virtual Private Network
XSS	Cross Site Scripting
WAF	Web Application Firewall
WHO	World Health Organization

Author details

Arun Warikoo
Cyber Security Specialist, Princeton, NJ, United States of America

*Address all correspondence to: arun.warikoo@gmail.com

References

[1] Bassett G, Hylender D, Langlois P, Pinto A, Widup S. Verizon Data Breach Investigation Report [Internet]. 2021. Available from: https://www.researchgate.net/publication/351637233_2021_Verizon_Data_Breach_Investigations_Report [Accessed: September 19, 2022]

[2] Sonica Wall. Cyber Threat Report: Mid-Year Update [Internet]. 2021. Available from: https://www.sonicwall.com/resources/white-papers/mid-year-2021-sonicwall-cyber-threat-report/ [Accessed: September 19, 2022]

[3] Aurangzeb S, Aleem M, Iqbal M, Islam A. Ransomware: A Survey and Trends [Internet]. 2017. Available from: https://www.researchgate.net/publication/317380115_Ransomware_A_Survey_and_Trends [Accessed: September 19, 2022]

[4] Cyber Big Game Hunting [Internet]. 2022. Available from: https://www.crowdstrike.com/cybersecurity-101/cyber-big-game-hunting [Accessed: September 19, 2022]

[5] Harford, I. The history and evolution of ransomware [Internet]. 2021. Available from: https://www.techtarget.com/searchsecurity/feature/The-history-and-evolution-of-ransomware [Accessed: September 16, 2022]

[6] Richardson R, North M. Ransomware: Evolution, Mitigation and Prevention. Vol. 4276. Faculty Publications; 2017. Available from: https://digitalcommons.kennesaw.edu/facpubs/4276

[7] Constantin L. Ryuk explained: Targeted, devastatingly effective ransomware [Internet]. 2021. Available from: https://www.csoonline.com/article/3541810/ryuk-explained-targeted-devastatingly-effective-ransomware.html [Accessed: September 16, 2022]

[8] Constantin L. REvil ransomware explained: A widespread extortion operation [Internet]. 2021. Available from: https://www.csoonline.com/article/3597298/revil-ransomware-explained-a-widespread-extortion-operation.html [Accessed: September 16, 2022]

[9] Myers L. LockBit 3.0 Ransomware Abuses Windows Defender to Load Cobalt Strike [Internet]. 2022. Available from: https://blogs.blackberry.com/en/2022/08/lockbit-3-0-ransomware-abuses-windows-defender-to-load-cobalt-strike [Accessed: September 19, 2022]

[10] Righi I. Ransomware in Q2 2022: Ransomware is Back in Business [Internet]. 2022. Available from: https://www.digitalshadows.com/blog-and-research/ransomware-in-q2-2022-ransomware-is-back-in-business/ [Accessed: September 19, 2022]

[11] Baker K. Ransomware as a Service (RaaS) Explained [Internet]. 2022. Available from: https://www.crowdstrike.com/cybersecurity-101/ransomware/ransomware-as-a-service-raas/ [Accessed: September 19, 2022]

[12] Trend Micro Research. Lockbit [Internet]. 2022. Available from: https://www.trendmicro.com/vinfo/us/security/news/ransomware-spotlight/ransomware-spotlight-lockbit [Accessed: September 13, 2022]

[13] Feeley B, Hartley B. Lunar Spider Sharing the Same Web [Internet].

2019. Available from: https://www.crowdstrike.com/blog/sin-ful-spiders-wizard-spider-and-lunar-spider-sharing-the-same-web/ [Accessed: September 17, 2022]

[14] Hanel A. Big Game Hunting with Ryuk: Another Lucrative Targeted Ransomware [Internet]. 2019. Available from: https://www.crowdstrike.com/blog/big-game-hunting-with-ryuk-another-lucrative-targeted-ransomware/ [Accessed: September 17, 2022]

[15] Sood K, Hurley S, Arsene L. Dark Side Goes Dark: How Crowd Strike Falcon Customers Were Protected [Internet]. 2021. Available from: https://www.crowdstrike.com/blog/falcon-protects-from-darkside-ransomware/ [Accessed: September 17, 2022]

[16] Osborne C. Dark Side explained: The ransomware group responsible for Colonial Pipeline attack [Internet]. 2019. Available from: https://www.zdnet.com/article/darkside-the-ransomware-group-responsible-for-colonial-pipeline-cyberattack-explained/ [Accessed: September 17, 2022]

[17] Meyers A. The Evolution of PINCHY SPIDER from Gand Crab to REvil [Internet]. 2021. Available from: https://www.crowdstrike.com/blog/the-evolution-of-revil-ransomware-and-pinchy-spider/ [Accessed: September 17, 2022]

[18] Elsad A, Gumarin JR, Barr A. LockBit 2.0: How This RaaS Operates and How to Protect Against It [Internet]. 2022. Available from: https://unit42.paloaltonetworks.com/lockbit-2-ransomware/ [Accessed: September 18, 2022]

[19] Elsad A. Threat Assessment: BlackByte Ransomware [Internet]. 2022. Available from: https://unit42.paloaltonetworks.com/blackbyte-ransomware/ [Accessed: September 19, 2022]

[20] Liska A, Gallo T. Ransomware. OReilly; 2016. pp. 53-55. ISBN: 978-1-491-96788-1

[21] Berrueta E, Morato D, Magana E, Izal M. A Survey on Detection Techniques for Cryptographic Ransomware. IEEE; 2019. DOI: 10.1109/ACCESS.2019.2945839

[22] Hills M, editor. Why Cyber Security Is a Socio-Technical Challenge: New Concepts and Practical Measures to Enhance Detection. Nova; 2016. p. 188. ISBN: 978-1-53610-090-7

[23] Hassan A. Ransomware Revealed. Apress; 2009. 209 p. DOI: 10.1007/9781484242551

Section 4

Mathematical Modeling

Chapter 6

A Review of Mathematical Model Based in Clustered Computer Network

Cristiane M. Batistela and José Roberto C. Piqueira

Abstract

The threats produced by viruses in computer networks have been frequent and the subject of many studies. Computer viruses share common characteristics with biological viruses, and therefore, one of the ways to study the dynamics of virus propagation has been through biological analogies. Inspired by macroscopic models, the susceptible-infected-removable (SIR) model allowed variations of compartmental models and suggested defenses considering antidotal (SIRA) and quarantined compartments (SIQRA), giving rise to models that evaluate the effectiveness and strategies to control the spread of viruses in networks. Recently, with the rapid popularization and access to networks, new studies have been taken into consideration the clusters of association of networks, indicating new control strategies and particularities of the dynamics. Toward this goal, this chapter presents a review of the mathematical model based in clustered computer network with the brief overview of the mathematical model reviews and providing an integrated framework to clustered model. In this essay, there is a discussion about the several ways of applying compartmental models to study the propagation of computer viruses and malwares through networks, emphasizing the effect of connections between geographically distributed machine clusters.

Keywords: bifurcation, cluster, disease-free, equilibrium point, SIR, stability

1. Introduction

Computer viruses emerged as programs capable of harming the functioning of a machine. Initially, the damage was minor as well as its proliferation capacity. With the increasing access to communication networks, the great development of hardware and software, and the inclusion of these services as an essential part of daily life, computer viruses have become a threat [1, 2].

Virus program codes are complex and easy to replicate, and detection and removal by antivirus programs are difficult [3]. Some feats of these viral programs are the ability to acquire bank passwords, personal data, and confidential information [4], which can cause immeasurable damage [5].

The increase in the use of mobile devices combined with the increasing access to wireless Internet facilitated the execution of many daily tasks such as accessing e-mail, electronic transactions of various natures and created opportunities for the advent of the Internet of Things (IoT), and connecting sensors and actuators to allow different types of objects to establish connections to the Internet, including home appliances, cars, and even industrial equipment. Therefore, these items are able to collect and transmit data from the cloud, contributing to a digital transformation in the world, and can provide several improvements in human life [6].

Consequently, understanding the spread of viruses in computer networks has become fundamental for the establishment of strategies to control and mitigate the spread of viruses. To improve the security and reliability of networks, a new branch of study, known as cybersecurity, has contributed to guide control strategies in order to minimize losses and one of its approaches is to build mathematical models.

One of the areas of cybersecurity is related to the study of the propagation of viruses in computer networks. Many ways of approaching the problem have been relevant to the understanding of the dissemination of malware, including the mathematical approach.

The mathematical study of computer viruses has an inspiration in biology and can be understood at two levels: microscopic and macroscopic [2, 7]. The tools used to develop antivirus programs, which are programs capable of detecting threats and preventing damage to machines, are concentrated at the microscopic level. In addition, the propagation of viruses in the network can be mitigated by the action of antivirus and quarantines proposed by the software when detecting some unexpected action [8].

The macroscopic level was developed from the classical model of disease propagation whose dynamics indicate the possibility of infection [9] and favor the orientation of strategies to control dissemination. The classic epidemiological model, proposed by Kermack and McKendrick, suggests the division of the population into compartments containing the group of susceptible (S), infected (I), and removed (R), giving rise to the SIR model, whose dynamics and parameters indicate strategies for control [9, 10].

Inspired by the above-mentioned works and based on the compartment level SIR model, this chapter considers the review of relationship between networks and the influence of the biolocical compartmental models for cybersecurity. Different from the conventional compartimental level models, this study shows the issue of how the association of two compartmental models occurs and indicates future prospects.

This work reviews the develop and analyze a model of virus propagation in two independent populations where those who ware infected from one population can come into contact with those who are susceptible from the other, to analyze the effects that one infected population can cause on the other. For this, two clusters were created and each cluster represents a population, both exposed to the same virus and represented by a model with antoidotal compartment. To represent the interaction between the two populations, a new connection between the sets was created and represents the capacity of an infected person come into contact with a susceptible one from the other, and this interaction will be modulated by a parameter.

The remaining chapter is ordered as follows: In the next section, a review of epidemiological models is presented with applications in computers, in section 3, hypothesis and equations are presented for the model with antidotal compartment, and the cluster model is presented, followed by the conclusion.

2. Epidemiological models

In 1927 was published for the first time a deterministic model to study the dynamics of virus spread in populations. This was a compartmental model consisting of three compartments (susceptible—infected—removed) [9]. In this work, a theory was developed relating the development of an epidemic to a critical value, later known as the basal reproduction number R_0.

The modeling of epidemics is associated with the dynamic behavior of processes where populations are studied according to their epidemiological status, these processes are described by differential equations, and the dynamics between their states is given by different parameters such as birth rate, mortality rate, infection, and recovery rate.

The modeling of the spread of epidemics has been the objective of many works [11–15]. These models allow a better understanding of the mechanisms of disease spread and can lead to more effective control strategies.

The scientific literature in epidemiology is quite diverse. Among the most cited models regarding this topic are the models: susceptible—infectious—susceptible (SIS) models [16–19]; susceptible—infected—recovered (SIR) models [20–25]; susceptible—exposed—infected—recovered (SEIR) models [26–28]; susceptible—exposed—infected—quarantined—recovered (SEIQR) models; susceptible—exposed—infected—quarantined—recovered—susceptible (SEIQRS) [29]; susceptible—exposed—infected—recovered—susceptible—vaccined (SEIRS-V) [30]; susceptible—exposed—infected—susceptible—vaccined (SEIS-V) [31] and others.

As for the way of treating chance, it can be classified into two levels: stocchastic and deterministic. In the first case, the model includes variables, giving a probabilistic distribution to the system, incorporating uncertainty, an intrinsic characteristic of epidemiological systems [32–35]. On the other hand, deterministic models provide the same results every time they are simulated with the same initial conditions [36, 37], being suitable to verify system sensitivity to the variation of the parameters [20, 21, 38].

Adapting the SIR model to computers, a lot of research has contributed to the understanding of virus propagation [26, 28–31, 39, 40] and one of the main goals is to establish effective security strategies [41].

The most explored strategies in cybersecurity are related to the use of antiviral compartments (A) and quarantine (Q). The adaptation of the SIR model gave rise to some robust models, including susceptible—infected—removed—antidotal (SIRA) [21] and susceptible—infected—removed—antidotal (SIQRA) [42].

Following this line, the first analysis of clustered computer networks using an epidemiological model studied the influence between two networks equipped with computers with antivirus and evaluated the dynamics of virus promotion and suggested viral dissemination control strategies.

3. Cluster SIRA model: Hypothesis and equations

There is a lot of compartimental models indicated for epidemiology [43] and their origin is Kermack and Mckendrick SIR (susceptible—infected—removed) models [9, 43, 44].

The population is considered constant and is divided into three compartments: Susceptible computers are uninfected and subject to infection (S); infected computers are represented by (I), those removed by infection or not (R), as shown in **Figure 1**.

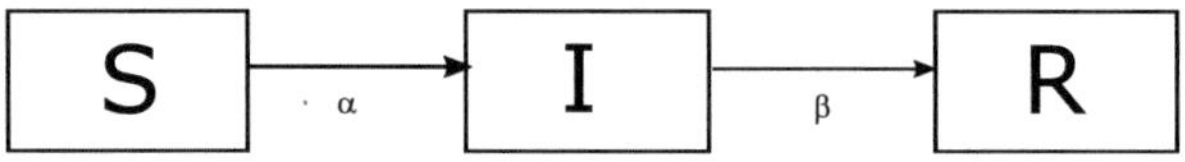

Figure 1.
SIR model.

As reported by [21] the dynamic equation for the populations S, I and R are:

$$\begin{cases} \dot{S} &= -\alpha SI; \\ \dot{I} &= \alpha SI - \beta I; \\ \dot{R} &= \beta I. \end{cases} \tag{1}$$

The susceptible population S is infected with a rate, that is, related to the probability of susceptible individuals to establish effective communications with infected ones. Therefore, this rate is proportional to the product SI, with proportion factor represented by α and infected individual can become removed with a rate controlled by β.

Considering initial conditions $S(0) \geq 0$, $I(0) \geq 0$ and $R(0) \geq 0$, in model such as SIR the interest is to investigate the dynamics of virus propagation indicates whether the virus will remain in the network or if it will naturally be eradicated. One of the ways to evaluate this behavior is to study the basal reproduction rate (R_0). This number indicates whether the virus will continue to be propagated and will be considered a situation analogous to the endemic one, or if it will become extinct in the network.

Based on a model described by (1), a model with a modification, including an antidotal population compartment (A) representing nodes of the network equipped with fully effective antivirus programs, is studied and considering constant population with four compartments: susceptible computers are uninfected and subject to infection (S); infected computers are represented by (I), and those removed by infection or not (R) and (A) are uninfected computers equipped with antivirus, as shown in **Figure 2**.

As reported by [21] the dynamic equation for the populations S, I, R and A are:

$$\begin{cases} \dot{S} &= N - \sigma_{SA}SA - \beta SI - \mu S + \sigma R; \\ \dot{I} &= \beta SI - \alpha_{IA}AI - \delta I - \mu I; \\ \dot{R} &= \delta I - \sigma R - \mu R; \\ \dot{A} &= \sigma_{SA}SA + \alpha_{IA}AI - \mu A. \end{cases} \tag{2}$$

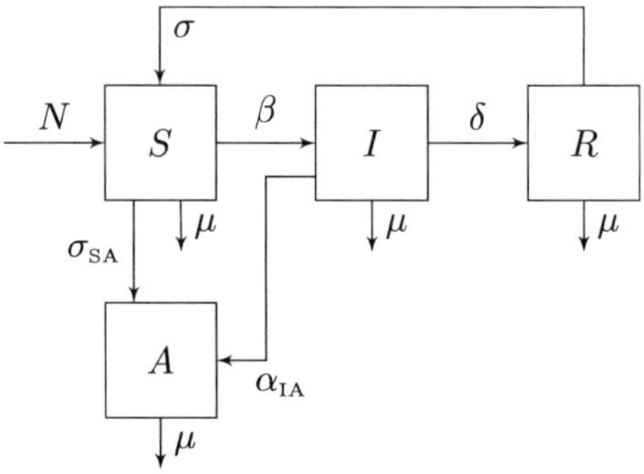

Figure 2.
SIRA model.

The influx rate is considered to be $N = 0$ because during the propagation of the considered virus, there is no incorporation of new computers in the network. The choice of $\mu = 0$ is justified that the machines obsolescence time is larger than the time of the virus action.

The model represents the spread of a known virus and the conversion of the antidoto to the infected is not considered. In this model, a vaccination strategy can be defined implying a control strategy associated with the economic use of antivirus programs.

The analysis of the SIRA model shows that it is possible to reach an disease-free equilibrium, guaranteeing a good operational performance of the network and that even in a situation of endemic equilibrium, the introduction of at least one machine equipped with antivirus guarantees a good performance of the network, tending to a disease-free equilibrium.

Furthermore, considering a constant total number of machines, the main control parameters are associated with the infection rate and how quickly infected machines are removed for formatting procedure. The other network parameters are associated with the transient response of the network in some small disturbance.

The other variation of the SIRA model is improved by considering that, when the machines pass to the removed condition, a fraction of these machines is recovered and the complement is considered dead. The introduction of the mortality rate results in an increase of the robustness of the disease-free equilibrium point of a computer network [38].

The study of the SIRA model was complemented by considering another control strategy by adding a quarantined compartment. The new compartment can be evaluated for the presence or absence of saturation and both situations indicate robustness in control strategies.

Based on a model described by [20], the virus propagation in a cluster is studied [45]. The model proposed is an association of two networks constituted by the SIRA model that interacts as shown in **Figure 3**.

Considering this hypothesis, adding another compartimental model, and associating a new infection rate, representing the infection capacity to the network, the cluster SIRA model for viruses propagation was proposed has the following dynamical eqs. (3):

$$\begin{cases} \dot{S}_1 &= -\alpha_{SA1}S_1A_1 - \beta_1 S_1 I_1 - \rho_2 I_2 S_1 + \theta 1 R_1; \\ \dot{I}_1 &= \beta_1 S_1 I_1 - \delta_1 I_1 - \alpha_{IA1} I_1 A_1 + \rho_2 I_2 S_1; \\ \dot{R}_1 &= \delta_1 I_1 - \theta_1 R_1; \\ \dot{A}_1 &= \alpha_{SA1} S_1 A_1 + \alpha_{IA1} I_1 A_1; \\ \dot{S}_2 &= \alpha_{SA2} S_2 A_2 - \beta_2 S_2 I_2 - \rho_1 I_1 S_2 + \theta 2 R_2; \\ \dot{I}_2 &= \beta_2 S_2 I_2 - \delta_2 I_2 - \alpha_{IA2} I_2 A_2 + \rho_1 I_1 S_2; \\ \dot{R}_2 &= \delta_2 I_2 - \theta_2 R_2; \\ \dot{A}_2 &= \alpha_{SA2} S_2 A_2 + \alpha_{IA2} I_2 A_2. \end{cases} \tag{3}$$

For the cluster SIRA model, the susceptible population S is infected with a rate, that is, related to the probability of susceptible elements to establish effective communications with infected ones and this rate is proportional to the product SI, with proportion factor represented by α or if infectivity occurs between network, by rate ρ that is related to the probability of infected elements to establish effective communications with susceptible computer in another network.

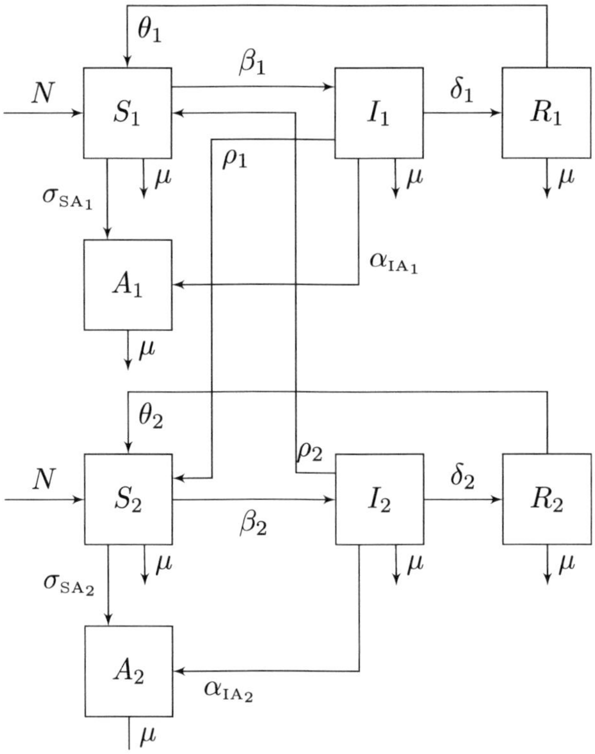

Figure 3.
SIRA cluster model.

Clustered sets, represented by subscripts 1 and 2, are divided into four groups, as shown in **Figure 3** and populations are considered constant in each cluster.

The SIRA cluster study, despite presenting a simple model composed of two connected grouped networks, points out the main control strategies associated with the control of parameters in order to avoid new forms of attacks.

Among the possibilities for adjustments, we consider infection rates in susceptible populations, due to contact with infected populations from the same cluster (β); conversion rates are removed by infected (δ) and infection rates in susceptible populations due to contact with the infected population of the other cluster (ρ).

The choice of network topology and connection strategies is an effective strategy to reduce the spread of viruses, since infection rates are not known in advance.

Another way to prevent the spread of viruses is to maintain the removal rates of damaged machines, which plays an important role in controlling the spread of networks.

If the virus is known, the best strategy to prevent its spread is to introduce antidote nodes containing programs that can be propagated throughout the network, immunizing the other nodes.

4. Conclusions

This chapter reports a detailed survey of compartments model in computer viruses with antidotal machine. The review provided a wide application of epidemiological models in compartmental models applied to computer networks. The focus of the review was to show the infuence of machines equipped with antivirus and their control strategies on the spread of viruses. The study of clustered networks is recent and with the model presented, it is expected that the review provides tools for studies of virus propagation dynamics in more complex networks.

Acknowledgements

JRCP is supported by the Brazilian Research Council (CNPq), grant number: 302883/2018–5.

Conflict of interest

The author declares that there is no conflict of interests regarding the publication of this study.

Author details

Cristiane M. Batistela[1] and José Roberto C. Piqueira[2]*

1 Federal University of ABC – UFABC, São Bernardo do Campo, SP, Brazil

2 Polytechnic School of University of São Paulo – EPUSP, São Paulo, SP, Brazil

*Address all correspondence to: piqueira@lac.usp.br

References

[1] Denning PJ, editor. Computers under Attack: Intruders, Worms, and Viruses (Vol. 990). New York: ACM Press; 1990

[2] Cohen F. Computer viruses: Theory and experiments. Computers & Security. 1987;**6**(1):22-35

[3] Tippett PS. The kinetics of computer virus replication: A theory and preliminary survey. In: Safe Computing: Proceedings of the Fourth Annual Computer Virus and Security Conference. 1991. pp. 66-87

[4] Yang LX, Yang X. A new epidemic model of computer viruses. Communications in Nonlinear Science and Numerical Simulation. 2014;**19**(6): 1935-1944

[5] Cohen FB. A Short Course on Computer Viruses. Pittsburgh, PA, USA: John Wiley & Sons, Inc; 1994

[6] Botta A, De Donato W, Persico V, Pescapé A. Integration of cloud computing and internet of things: A survey. Future Generation Computer Systems. 2016;**56**:684-700

[7] Kephart JO. Direct-graph epidemiological models of computer virus. In: Proceedings of IEEE Symposium on Security and Privacy. Oakland, CA, USA: IEEE; 1991

[8] Kephart JO, White SR. Measuring and modeling computer virus prevalence. In: Proceedings 1993 IEEE Computer Society Symposium on Research in Security and Privacy. Oakland, CA, USA: IEEE; 1993. pp. 2-15

[9] Kermack WO, McKendrick AG. Contributions of mathematical theory to epidemics. Proceedings of the Royal Society of London, Series A, Containing Papers of a Mathematical and Physical Character. 1927;**115**(772):700-721

[10] Kermack WO, McKendrick AG. Contributions to the mathematical theory of epidemics. II.—The problem of endemicity. Proceedings of the Royal Society of London. Series A, Containing Papers of a Mathematical and Physical Character. 1932;**138**(834):55-83

[11] Anderson RM, Anderson B, May RM. Infectious Diseases of Humans: Dynamics and Control. New York, USA: Oxford University Press; 1992

[12] Murray JD. Mathematical Biology. 3rd ed. New York: Springer-Verlag; 2002

[13] Clancy D. Optimal intervention for epidemic models with general infection and removal rate functions. Journal of Mathematical Biology. 1999;**39**(4): 309-331

[14] Allen LJ, Brauer F, Van den Driessche P, Wu J. Mathematical Epidemiology. Vol. 1945. Berlin: Springer; 2008. p. 2008

[15] Brauer F, Castillo-Chavez C, Castillo-Chavez C. Mathematical Models in Population Biology and Epidemiology. Vol. 2. New York: Springer; 2012. p. 2012

[16] Amador J, Artalejo JR. Modeling computer virus with the BSDE approach. Computer Networks. 2012;**57**(1):302-316

[17] Sanders J, Noble B, Van Gorder RA, Riggs C. Mobility matrix evolution for an SIS epidemic patch model. Physica A; Statistical Mechanics and its Applications. 2012;**391**(24):6256-6267

[18] Wang Y, Cao J, Jin Z, Zhang H, Sun GQ. Impact of media coverage on

epidemic spreading in complex networks. Physica A; Statistical Mechanics and its Applications. 2013; **392**(23):5824-5835

[19] Tomovski I, Trpevski I, Kocarev L. Topology independent SIS process: An engineering viewpoint. Communications in Nonlinear Science. 2014;**19**(3):627-637

[20] Piqueira JRC, De Vasconcelos AA, Gabriel CE, Araujo VO. Dynamic models for computer viruses. Computers and Security. 2008;**27**(7–8):355-359

[21] Piqueira JRC, Araujo VO. A modified epidemiological model for computer viruses. Applied Mathematics and Computation. 2009;**213**(2):355-360

[22] Ren J, Yang X, Yang LX, Xu Y, Yang F. A delayed computer virus propagation model and its dynamics. Chaos Soliton & Fractals. 2012;**45**(1): 74-79

[23] Wierman JC, Marchette DJ. Modeling computer virus prevalence with a susceptible-infected-susceptible model with reintroduction. Computational Statistics & Data Analysis. 2004;**45**(1):3-23

[24] Zhu Q, Yang X, Ren J. Modeling and analysis of the spread of computer virus. Communications in Nonlinear Science. 2012;**17**(12):5117-5124

[25] Shukla JB, Singh G, Shukla P, Tripathi A. Modeling and analysis of the effects of antivirus software on an infected computer network. Applied Mathemaics and Computation. 2014; **227**:11-18

[26] Mishra BK, Saini DK. SEIRS epidemic model with delay for transmission of malicious objects in computer network. Applied Mathematics and Computation. 2007; **188**(2):1476-1482

[27] Wang F, Zhang Y, Wang C, Ma J. Stability analysis of an e-SEIAR model with point-to-group worm propagation. Communications in Nonlinear Science. 2015;**20**(3):897-904

[28] Mishra BK, Pandey SK. Dynamic model of worms with vertical transmission in computer network. Applied Mathematics and Computation. 2011;**217**(21):8438-8446

[29] Mishra BK, Jha N. SEIQS model for the transmission of malicious objects in computer network. Applied Mathematical Modelling. 2010;**34**(3): 710-715

[30] Mishra BK, Keshri N. Mathematical model on the transmission of worms in wireless sensor network. Applied Mathematical Modelling. 2013;**37**(6): 4103-4111

[31] Mishra BK, Pandey SK. Dynamic model of worm propagation in computer network. Applied Mathematical Modelling. 2014;**38**(7–8):2173-2179

[32] Radha M, Balamuralitharan S, Geethamalini S, Geetha V, Rathinasamy A. Analytic solutions of the stochastic SEIA worm model by homotopy perturbation method. AIP Conference Proceedings. 2019;**2112**(1): 020050

[33] Amador J, Artalejo JR. Stochastic modeling of computer virus spreading with warning signals. Journal of the Franklin Institute. 2013;**350**(5):1112-1138

[34] Amador J. The stochastic SIRA model for computer viruses. Applied Mathematics and Computation. 2014; **232**:1112-1124

[35] Zhang C, Zhao Y, Wu Y, Deng S. A stochastic dynamic model of computer viruses. Discrete Dynamics in Nature and Society. 2012;**2012**:1-16

[36] Geethamalini S, Balamuralitharan S, Radha M, Geetha V, Rathinasamy A. Stability analysis of deterministic SEIA worm model by reproductive number. AIP Conference Proceedings. 2019; **2112**(1):020044

[37] Geetha V, Balamuralitharan S, Geethamalini S, Radha M, Rathinasamy A. Analytic solutions of the deterministic SEIA worm model by homotopy perturbation method. AIP Conference Proceedings. 2019;**2112**(1): 020100

[38] Batistela CM, Piqueira JRC. SIRA computer viruses propagation model: Mortality and robustness. International Journal of Applied and Computational Mathematics. 2018;**4**(5):128

[39] Martcheva M. An Introduction to Mathematical Epidemiology. Vol. 61. New York: Springer; 2015. p. 2015

[40] Wang F, Zhang Y, Wang C, Ma J. Stability analysis of an e-SEIAR model with point-to-group worm propagatio. Communications in Nonlinear Science. 2015;**20**(3):897-904

[41] Li P, Yang X, Xiong Q, Wen QJ, Tang YY. Defending against the Advanced Persistent Threat: An Optimal Control Approach. Security and Communication Networks. 2018; **2018**:1-14

[42] Piqueira JRC, Batistela CM. Considering quarantine in the SIRA malware propagation model. Mathematical Problems in Engineering. 2019

[43] Kermack WO, McKendrick AG. Contributions of mathematical theory to epidemics. Proceedings of the Royal Society of London, Series A, Containing Papers of a Mathematical and Physical Character. 1932;**138**(834):55-83

[44] Kermack WO, McKendrick AG. Contributions to the mathematical theory of epidemics - further studies of the problem of endemicity. Proceedings of the Royal Society of London, Series A, Containing Papers of a Mathematical and Physical Character. 1933;**141**(843): 94-122

[45] Piqueira JRC, Cabrera MA, Batistela CM. Malware propagation in clustered computer networks. Physica A: Statistical Mechanics and its Applications. 2021;**573**:125958